BLUE GOLDFISH 2.0

8 WAYS TO LEVERAGE AI AND DATA FOR CUSTOMER LOYALTY & MARKETING INNOVATION

STAN PHELPS
&
CHUCK GALLAGHER

Blue Goldfish 2.0 is a second edition update to the original *Blue Goldfish*. The original was published in 2016 by Stan Phelps and Evan Carroll.

Published by 9 INCH

ISBN: 978-1-952234-15-6
1. Artificial Intelligence 2. Analytics 3. Data 4. Customer Experience

First Printing: 2024
Printed in the United States of America

Blue Goldfish 2.0 is available for bulk orders. For further details and special pricing, please e-mail stan@purplegoldfish.com

"This book is dedicated to my two teenage sons,
Thomas & James."

\- Stan Phelps

"This book is dedicated to my youngest son, Alex,
who opened my eyes to the power of AI."

\- Chuck Gallagher

TABLE OF CONTENTS

FOREWORD

BY SHEP HYKEN

In today's business landscape, the conversation around artificial intelligence (AI) is impossible to ignore. It permeates every aspect of industry discussions, reshaping how businesses operate and engage with customers. AI, at its core, harnesses the power of data to simulate human intelligence, enabling machines to learn, reason, and make decisions autonomously. While the concept of AI encompasses more than just data processing, its application in customer experience is profound and transformative.

As Evan Carroll aptly phrases it, "Bigger, so the establishment believes, is better." However, an alternative perspective on AI is emerging among forward-thinking companies.

In contrast to the traditional focus on 'big data,' there's a growing recognition of the significance of 'little data'—the personalized insights gleaned from individual interactions and behaviors. Rather than solely relying on massive datasets to identify broad trends, companies are leveraging AI to decipher subtle nuances and preferences unique to each customer. These insights, powered by AI algorithms and machine learning models, underpin what authors Phelps and Gallagher term the Blue Goldfish concept. For instance, AI algorithms can predict a customer's preferences based on past interactions, enabling businesses to tailor experiences with unprecedented precision.

AI enables businesses to not only understand customer behavior but also anticipate their needs and preferences in real-time. It's not just about processing data; it's about harnessing AI to enhance the human element of customer interactions, creating meaningful connections that resonate on a personal level.

In essence, while many companies focus solely on the breadth of data through AI, it's the depth of insights derived from AI-driven personalization that truly sets them apart. The most successful companies leverage AI to cultivate relationships, responsiveness, and readiness through meticulous attention to detail. They create Blue Goldfish experiences that not only drive profits but also foster prophetic insights.

This book presents an invigorating challenge to your organization. By its conclusion, you'll confront a pivotal choice: continue fixating on AI-driven data trends or embrace the transformative potential of AI-driven personalization. You'll soon realize that in the realm of customer experience, it's not just about the scale of data processed by AI—it's about the depth of insights and the quality of interactions that matter most. And once you embark on this journey, you'll inevitably ask yourself, "How can AI elevate the experience we provide and what's our Blue Goldfish?"

Shep Hyken, CSP, CPAE (www.hyken.com) is a customer service expert, professional speaker, and New York Times bestselling business author.

INTRODUCTION

BY STAN PHELPS

"There are only two sources of competitive advantage: the ability to learn more about our customers faster than the competition and the ability to turn that learning into action faster than the competition."

– Jack Welch

Each year autumn rolls around. Summer tans have faded. Marketing planning begins. Here's the sequence: Agencies get briefed, ideas are presented, concepts get developed, creative gets shot, media gets bought, ads run, results get measured, and (possibly) awards get submitted. Next up is a celebratory dinner with the requisite toasts, back slapping, and high-fives. Autumn rolls around again. Lather, rinse, repeat.

This is old school traditional marketing. A phenomenon that Seth Godin labels the TV-industrial complex[1]. The only problem is that it doesn't work as well as it used to. Or in some cases it doesn't work at all.

Who's to blame for the campaign-based mentality? Is it the brand or is it the agency? Who is going to break the cycle of T.I.T.W.W.A.D.I. (This Is The Way We've Always Done It)? It's hard to blame the agency. You can't expect them to change if their entire business model is predicated on not doing so. Perhaps it's time to break the cycle.

---

1. http://sethgodin.typepad.com/seths_blog/2004/04/the_end_of_mass.html

Are there companies willing to change? In a *Forbes* interview, former CVS marketing chief Rob Price describes this shift in focus:

> We've gone through a real transition. We've moved from being more of an Advertising and Promotions department to really being an Insights to Action function. That changes your focus substantially. But of course, we still do the advertising, marketing, and promotional accountability; that's our functional expertise... We've reduced the degree of traditional advertising that we've done. Because we've found, and maybe it stems from our advertising roots. Advertising is about frequency and reach. Well, we have 35 million people coming into stores every week. Tens of millions with our mail-order pharmacy and with our prescription benefit management. So, if you monetized, if you put into traditional advertising terms, all of those are exposures or impressions. The media in our operating environment is thousands of stores, hundreds of thousands of people, serving millions of customers, creating billions of interactions. It's very large numbers, and we've decided to harness the energy of the marketing team in collaboration with IT, Store Operations, and Digital[2].

CVS is shifting focus away from awareness and acquisition. Marketing is no longer a one-to-many function and the race is on. Companies are now faced with the challenge of using technology, data, and analytics to focus on many-to-one marketing and AI is at the forefront of accelerating this process.

---

2. http://www.forbes.com/sites/onmarketing/2014/03/05/the-future-of-marketing-a-little-less-campaign-and-a-little-more-action/

KNOWLEDGE AND AGILITY

Customer experience has become new battleground for competitive advantage. But with every wise company focusing on customer experience, how do you stand out?

The late Jack Welch perhaps said it best, "There are only two sources of competitive advantage: the ability to learn more about our customers faster than the competition and the ability to turn that learning into action faster than the competition."

Competitive advantage then must come from either knowledge or agility. That is, what you know about your customers or how quickly you can put that into action.

Imagine what would happen if your competitive advantage could come from both knowledge and agility. This is where AI, data, and analytics come into play.

INFO-SENSE AND CUSTOMER EXPERIENCE

Info-sense is "the ability to utilize data to really understand customers as people and personalize their service." Coined by TrendWatching.com, this sixth-sense capability will enable companies to learn more about customer's individual needs and put that insight into action in real time.

Companies who embrace info-sense can provide customers with exactly what they need right when they need it.

The shift to info-sense should be good news for business. The benefit is simple. It typically costs one-tenth the amount of money to upsell a current customer that it does to acquire a new one.

In today's digital ecosystem, the relevance of info-sense is more pronounced than ever. Here's why:

1. **Customer Expectations**: Consumers increasingly expect personalized experiences. They desire services and products that resonate with their individual needs and preferences. Info-sense enables businesses to meet these expectations by leveraging data to tailor their offerings and interactions.
2. **Competitive Advantage**: In a saturated market, personalization becomes a key differentiator. Businesses that effectively utilize info-sense can distinguish themselves by offering uniquely relevant experiences to their customers, thereby fostering loyalty and competitive advantage.
3. **Technological Advancements**: The proliferation of data analytics tools and AI-driven technologies has made it easier for businesses to implement info-sense in their operations. These technologies can process vast amounts of data in real-time, providing actionable insights to enhance customer experiences.
4. **Economic Efficiency**: Personalized experiences not only enhance customer satisfaction but can also lead to more efficient marketing and sales strategies. By understanding customer needs better, businesses can allocate resources more effectively, reducing waste and increasing ROI.

Info-sense is not a fleeting trend but a foundational element of modern business strategy that aligns with the ethos of customer-centricity and technological innovation. As businesses navigate the complexities of the digital age, the principles underpinning info-sense offer a roadmap for creating meaningful, personalized, and ethical customer engagements that resonate in today's market and beyond.

THE POWER OF INSIGHTS TO ACTIONS

Want an example of a brand driving business through info-sense? Enter Westpac. An Australian financial-services provider, Westpac is one of Australia's big four banks and the second-largest bank in New Zealand. With over 12 million customers, they are approaching 100 million customer interactions per month. The bank has been able to connect the dots and embrace info-sense. They leveraged their immense amount of customer data with a program called "*Know Me.*"

According to Westpac's former GM of Customer Relationship Marketing Karen Ganschow, the bank knows needs are driven by life stages and moments of truth in their customer journey. The bank uses analytics to drive insights about their customers. These insights allow them to create NBO's (next best offers). By using a mix of offline transactional data and online behavioral data, the insights allow the bank to predict what the customer might need next. Standing NBOs are placed into Westpac's front-line systems and the next time a customer interacts with a bank teller or a customer service representative, the employee extends the offer.

Despite high confidence in these personalized offers, Westpac knows a 100 percent close rate is unreasonable. They also understand that a one percent conversion creates a poor customer experience. In the first year of the "Know Me" program, Westpac converted 46 percent of offers. There's no doubt that using big data trends and little data insights helped identify a range of personalized "next best offers" and ultimately delivered millions to Westpac's bottom line.

Blue Goldfish 2.0 is about helping you do the same for your organization. It follows the original *Blue Goldfish*, which I coauthored with Evan Carroll in 2016. Why Blue and Why a Goldfish? That's coming up next.

PART I:

WHAT IS A BLUE GOLDFISH?

CHAPTER 1

THE AMAZON EFFECT

"The balance of power is shifting toward consumers and away from companies... The right way to respond to this if you are a company is to put the vast majority of your energy, attention, and dollars into building a great product or service and put a smaller amount into shouting about it, marketing it."

– Jeff Bezos

Customer expectations are binary. You either exceed them, or you fall short. Similarly, if you're trying to keep up with increasing consumer expectations, you're fighting an uphill battle.

Amazon, no longer a small online bookstore by any measure, knows this. Their objective isn't to meet expectations. Rather, their objective is to exceed customer expectations such that customer expectations increase through the service they provide. Scot Wingo, Co-Founder of ChannelAdvisor, calls this increasing customer expectation the "Amazon Effect."

Described as a complex and powerful force, The Amazon Effect shows how much digital innovation and tech advancements can shake up the usual ways of doing business and how we shop. It's a wake-up call for businesses to quickly evolve or risk falling behind in a world where digital savvy and putting customers first are becoming the new norms.

THE EMPOWERED ON-DEMAND CUSTOMER

Fueled largely by the Amazon Effect, the expectations of today's customer are rising at a dramatic pace. They expect a personalized, quick, and consistent experience. An IBM Institute for Business Value report brings this fact into sharp focus:

- 76 percent of consumers expect organizations to understand their individual needs.
- 81 percent of consumers demand improved response times.
- 68 percent anticipate organizations will harmonize consumer experiences.

Empowered customers are starting to take for granted that a company will know and understand their individual needs. They expect companies to know what they searched for and their past ordering history. Amazon has raised the bar on customer expectations.

David Trice of ENGAGEcx refers to this new breed of need-it-now consumers as On-Demand customers. Today's customer expects their needs will be addressed with precision and expertise at every touchpoint. Companies that leverage info-sense can expect increased customer loyalty and advocacy. The penalty for non-performers is more than just a lost customer as social media now provides a platform for any unhappy consumer to broadcast their bad experience globally.

To offset the Amazon Effect, you must get ahead of the expectations curve. You don't have to deliver Amazon-level service, but you must know that expectations are increasing. And you must stay one step ahead.

The Amazon Effect continues to be a driving force in the retail landscape, shaping not just consumer behavior but also the strategic direction of businesses worldwide. This phenomenon, rooted in Amazon's massive e-commerce influence, has set unprecedented standards in customer service, supply chain efficiency, and data-driven personalization. As we move forward, new trends are emerging under its shadow. The integration of AI and machine learning in retail operations offers unparalleled customer insights, leading to more personalized and engaging shopping experiences. The rise of social commerce signifies a shift towards more interactive and socially integrated purchasing journeys. Moreover, there's a growing emphasis on ethical consumerism and sustainability, areas where Amazon's vast reach and influence could either foster positive change or present new challenges. This evolution calls for businesses to not only adapt to these trends but also critically assess their impact on society and the environment. As Amazon paves the way, it's vital for companies to balance technological advancement with ethical considerations, ensuring a future where innovation and responsibility go hand in hand.

THE SECOND MACHINE AGE

We are entering a new era driven by AI. A time that MIT's Erik Brynjolfsson and Andrew McAfee call the Second Machine Age. We are now on the cusp of a time where business can start to leverage the advances in computer processing, artificial intelligence, and networked communication. A great example of this is Waze. The app allows users to access maps and driving directions. It also provides up to date conditions, allowing the driver to navigate swiftly and safely. It can do this not only because each road map has been scanned into the database, but because every cell phone transmitting from every car will reveal where the traffic jams are in real time.

Waze leverages crowdsourced data from its users, which allows it to provide real-time traffic updates and route suggestions. This dynamic data collection method is more adaptable and timely compared to traditional navigation systems that rely on less frequent updates. Real time data collection is rapidly changing the face of technology.

Waze now leverages AI in several ways to continually improve its service. It leverages each of the Blue Goldfish 3 R's:

1. **Relationship**: Waze employs machine learning algorithms to analyze user behavior patterns. By understanding how users interact with the app, such as their preferred routes, common destinations, and typical travel times, Waze can personalize its recommendations and notifications to provide more relevant and timely information to users.

2. **Responsiveness:** Waze uses machine learning algorithms to improve its estimated time of arrival predictions (ETAs). By considering factors such as historical traffic data, road conditions, weather forecasts, and even events like concerts or sports games, Waze can provide more accurate ETAs tailored to individual users' routes and preferences.
3. **Readiness:** Waze fosters a strong community of users who contribute valuable data and insights to the platform. Through features like reporting accidents, hazards, and police sightings, users help improve the accuracy of Waze's information. AI algorithms analyze and prioritize these reports to provide real-time updates to all users, enhancing safety on the road.

Overall, Waze's use of AI enables it to deliver a more personalized and responsive navigation experience, helping users save time and avoid traffic congestion while also contributing to a broader community-driven effort to improve road safety and efficiency.

It took several decades for earlier breakthrough technologies, such as the steam engine or electricity, to reach the point of ubiquity and flexible application in the first machine age. Once they did, both fundamentally changed the way people lived and businesses operated. Information technology and digital communication are now just reaching that same inflection point.

The big winners in this new era will be consumers, who will be able to buy a wider range of higher-quality goods and services at lower prices. The other winners will be those who create and finance the infrastructure or figure out how best to use them to gain competitive advantage. Great wealth will be created in the process[3].

3. https://www.washingtonpost.com/opinions/review-the-second-machine-age-by-erik-brynjolfsson-and-andrew-mcafee/2014/01/17/ace0611a-718c-11e3-8b3f-b1666705ca3b_story.html

WE DIDN'T START THE FIRE

We Didn't Start the Fire is a song by Billy Joel. Billy came up with the idea when he was turning 40. A chance meeting with Sean Lennon (son of John and Yoko) made him realize how each subsequent generation feels things are changing and getting worse.

The 1989 song lists 117 historical events from the 1940s to the late 80s, chronicling everything from scientific advances to political events in chronological order. We will do a similar exercise. We didn't get to this second machine age inflection point without tremendous innovation over the last 60+ years.

Let's look at some of the major breakthroughs over that time starting in the early 1960s:

1961 - The silicon chip was invented by two American electrical engineers, Jack Kilby and Robert Noyce. Their creation revolutionized and miniaturized technology and paved the way for the development of the modern computer. Until the chip was invented, most electrical devices were constructed using bulky, power-hungry vacuum tube technology. The development of transistors partially solved the problem, but these still had to be wired to circuit boards. Kilby and Noyce hit on the solution almost simultaneously, combining separate components in an integrated circuit made of a semiconductor material. Intel founder Robert Noyce, working in Palo Alto, California, favored silicon and can thus be credited as the person who put the Silicon into "Silicon Valley.[4]"

1964 - Roger Easton begins experiments that lead to the development of GPS (global positioning system), a ubiquitous feature of modern life. What began as a way of tracking the location of the

4. http://edition.cnn.com/2004/TECH/12/27/explorers.silicon/index.html

Russian Sputnik 1 satellite became a way for satellites to track us here on the surface of Earth[5].

1967 - Graduate student James Goodnight joins a project at North Carolina State University. The goal was to create a statistical analysis system (SAS) used by agricultural departments at colleges and universities. In 1976, SAS Institute becomes an independent, private business led by Dr. Goodnight.

1969 - Neil Armstrong becomes the first person to step onto the lunar surface of the moon. The Apollo missions would spawn technological advances such as water purification, breathing masks, polymer fabric, and cordless devices.

1969 - ARPANET (Advanced Research Projects Agency Network) headed by J.C.R. Licklider is launched. Licklider's idea of an intergalactic computer network, where everyone on the globe is interconnected with the ability to access programs and data at any site, from anywhere, begins to take shape[6].

1970 - Gordon Moore coins Moore's Law, a computing term that states that processor speeds, or overall processing power (the number of transistors on a CPU) for computers will double every two years[7]. Experts agree that computers should reach the physical limits of Moore's Law at some point this decade. While the strict interpretation of Moore's Law may be nearing its limits, ongoing research and development efforts in semiconductor technology will likely lead to continued advancements in computing power and performance through alternative approaches and innovations.

---

5. http://www.theatlantic.com/technology/archive/2014/06/the-man-who-created-gps/372846/
6. http://www.computerweekly.com/feature/A-history-of-cloud-computing
7. http://www.mooreslaw.org/

1973 - The Xerox Alto personal computer is developed at Xerox PARC. It becomes the first computer to use the desktop metaphor and mouse-driven graphical user interface (GUI).

1973 - The world's first mobile phone call was made when Martin Cooper, a senior engineer at Motorola, called a rival telecommunications company and informed them he was speaking via a mobile phone. The phone Cooper used, if you could call it that, weighed a staggering 2.5 pounds (1.1 kilograms). With this large prototype, you got 30 minutes of talk-time and it took 10 hours to charge[8].

1975 - The first personal computers were introduced. The MITS Altair 8800 was followed by the IMSAI 8080, an Altair clone. Bill Gates and Paul Allen wrote a BASIC compiler for the Altair and formed Microsoft.

1976 - Steven Sasson, a 24-year-old engineer at Eastman Kodak, invents the process to make digital photos.

1976 - A couple of other Steve's, Jobs and Wozniak, launch the Apple I. As Apple's only "kit" computer, consumers need to add a keyboard, power supply, and enclosure to the assembled motherboard around the 6502 processor.

1978 - Dan Bricklin and Bob Frankston introduce VisiCalc. The first electronic spreadsheet, it turned the personal computer into a useful business tool, not just a game machine or replacement for the electric typewriter.

1982 - The first thing (as in internet of things) connected to the internet was a Coca Cola vending machine at Carnegie Mellon University. Programmers connected to the machine over the Internet, allowing them to check the status of the machine and determined

8. http://www.knowyourmobile.com/nokia/nokia-3310/19848/history-mobile-phones-1973-2008-handsets-made-it-all-happen

whether there would be a cold Coke awaiting them should they decide to make the trip downstairs to the machine[9].

1983 - On a boat trip to Catalina Island in 1983, five marketing professionals conceived an electronic alternative that would transform the retail landscape. The solution involved using grocery scanners to distribute targeted coupons based on your purchases. Catalina Marketing was born. Catalina created a single solution–the Checkout Coupon–benefitting retailers, brands, and consumers[10].

1983 - Motorola releases its first commercial mobile phone, known as the Motorola DynaTAC 8000X. The handset offered 30 minutes of talk-time, six hours standby, and could store 30 phone numbers. It cost nearly $4,000[11].

1989 - Tim Berners-Lee, a British scientist at CERN, invents the World Wide Web. The web was originally conceived and developed to meet the demand for automatic information-sharing between scientists in universities and institutes around the world. The first website at CERN was dedicated to the World Wide Web project itself and was hosted on Berners-Lee's NeXT computer. The website described the basic features of the web; how to access other people's documents and how to set up your own server.

1990 - Quantum Computer Services, Inc., introduces Promenade, an online service that will be offered with the IBM PS/1 computer. home computer has been introduced with a built-in modem and online services to provide families immediate access to live, interactive education and entertainment services. Quantum Computer Services eventually became America Online[12].

---

9. http://whatis.techtarget.com/definition/Internet-of-Things
10. https://www.catalina.com/company/about-us
11. http://www.knowyourmobile.com/nokia/nokia-3310/19848/history-mobile-phones-1973-2008-handsets-made-it-all-happen
12. http://www.technologizer.com/2010/05/24/aol-anniversary/

1991 - Linus Torvalds invents Linux. Torvalds, a student at the University of Helsinki in Finland, begins writing his own kernel. He started by developing device drivers and hard-drive access, and by September had a basic design that he called Version 0.01. This kernel, which is called Linux, was afterwards combined with the GNU system to produce a complete free operating system[13].

1992 - Neil Papworth, a 22-year-old test engineer for Sema Group (now Airwide Solutions), sent the first text message on December 3, 1992, from his personal computer to the Vodafone network to the phone of Richard Jarvis. The text message read "Merry Christmas." AT&T would be the first to offer the service in the US in 2000[14].

1993 - CERN puts an updated version of World Wide Web software into the public domain. CERN makes the release available with an open license to maximize its dissemination. Through these actions, making the software required to run a web server freely available, along with a basic browser and a library of code, allowed the web to flourish.

1994 - Katie Couric and Bryant Gumbel infamously stumble when confronted with their first email address on *The Today Show*. Bryant Gumbel asks, "What is internet, anyway?"

1994 - Jeff Bezos incorporates the company Cadabra on July 5th. A year later he changes the name to Amazon after a lawyer misheard its original name as "Cadaver."

1994 - Netscape made its first internet browser available on October 13th. This browser, known as Netscape Navigator, played a pivotal role in popularizing the World Wide Web among the

13. http://www.livinginternet.com/i/iw_unix_gnulinux.htm
14. http://www.edn.com/electronics-blogs/edn-moments/4402146/1st-text-message-is-sent--December-3--1992

general public. It was one of the first browsers to be widely used and greatly contributed to the internet boom of the 1990s. Netscape Navigator's user-friendly interface and enhanced features, such as supporting graphics and multimedia, made it a significant tool in the expansion and accessibility of the internet during its early years.

1995 - Amazon.com goes online.

1995 - Pierre Omidyar auctions off a single broken laser pointer via a site he'd developed called AuctionWeb. Would it sell? To his surprise it sold for $14.83 and an idea was formed. The company was soon renamed eBay, short for Echo Bay, the name of Omidyar's consulting firm. eBay's vision for success transitioned from one of commerce—buying and selling things—to become something far more significant. It thrived because it placed customer data being at the heart of the business from its inception. eBay democratized ecommerce. The site pioneered an ecommerce marketing revolution, particularly when it comes to utilizing customer data to deliver a more engaging, seamless online experience. From day one they've used customer data to improve platform experience and they're only getting better at contextualizing their data sources. Its lesson to the internet has been that success comes from continual data innovation.

1995 - Google begins as a research project by Larry Page and Sergey Brin. Both are Ph.D. students at Stanford University.

1999 - Kevin Ashton coins the term "the Internet of Things" (IoT) while working at Auto-ID Labs[15].

1999 - Oracle executive Marc Benioff invites three friends to his San Francisco apartment. His business idea gets a lukewarm response. Co-founder Dave Moellenhoff doesn't sugarcoat it,

15. http://www.theguardian.com/media-network/2015/mar/31/the-internet-of-things-is-revolutionising-our-lives-but-standards-are-a-must

"You're an idiot. That's the stupidest thing. This is never going to work." The group presses forward and launches Salesforce, one of the first enterprise cloud software services in the world. The company pioneered the concept of delivering enterprise applications via a simple website[16].

2000 - Confinity and Elon Musk's X.com merge in March. The combined company becomes PayPal to sync up the name of the company with the name of the product.

2000 - The United States stops intentionally degrading GPS signals available to the public. Originally developed by the Department of Defense to aid the military, the satellite-based system provides location and time data to users. In announcing the discontinuation of the feature that deliberately degraded the signal, the White House said in a statement that civilian users of GPS would be able to pinpoint locations up to 10 times more accurately than before[17].

2001 - A free user-generated online encyclopedia called Wikipedia comes online and quickly becomes the reference site of choice for Internet users[18].

2002 - Amazon launches Amazon Web Services, a suite of cloud-based services including storage, computation, and even human intelligence through the Amazon Mechanical Turk[19].

2003 - 500 million devices are connected to the Internet. One device for every 12 people on the planet.

16. http://www.computerweekly.com/feature/A-history-of-cloud-computing
17. http://abcnews.go.com/Technology/AheadoftheCurve/top-10-innovations-decade/story?id=9204931
18. http://abcnews.go.com/Technology/AheadoftheCurve/top-10-innovations-decade/story?id=9204931
19. http://www.computerweekly.com/feature/A-history-of-cloud-computing

2003 - Chris Stolte and Pat Hanrahan invent a technology called VizQL at Stanford. VizQL attacks a basic business problem: making databases and spreadsheets understandable to ordinary people. Now, visualization is part of the journey and not just the destination. Fast analytics and visualization for everyone becomes the basis for the founding of Tableau[20].

2003 - Scientists announced that they had sequenced the entire human genome two years ahead of schedule. The 13-year international project set out to identify the 20,000 to 25,000 genes in human DNA[21].

2004 - 19-year-old Mark Zuckerberg launches thefacebook.com in his dorm room as a Harvard University sophomore.

2005 - YouTube is founded by former PayPal employees Chad Hurley, Steve Chen, and Jawed Karim. The idea was born at a dinner party in San Francisco the year before. The inspiration was two key events: Janet Jackson's wardrobe malfunction at the Super Bowl and a devastating tsunami in the Indian Ocean[22].

2006 - Doug Cutting begins working at Yahoo. They spin out the storage and processing parts of Nutch to form Hadoop as an open source Apache Software Foundation project. Why Hadoop? It was named after Cutting's son's stuffed elephant.

2007 - The first iPhone was released combining the internet, a cellphone, and an iPod into one device. In 74 days, Apple sells one million iPhones.

---

20. https://www.tableau.com/about/mission-values
21. http://abcnews.go.com/Technology/AheadoftheCurve/top-10-innovations-decade/story?id=9204931
22. http://www.businessinsider.com/key-turning-points-history-of-youtube-2013-2?op=1

2007 - Waze is founded. The Israeli and Palo Alto-based company is a developer of free mapping and turn-by-turn navigation apps for iOS and Android devices.

2007 - Sitting in Zeke's Sports Bar, Tom Preston-Werner shares an idea with Chris Wanstrath, "I launched into my half-baked idea for some sort of website that acted as hub for coders to share their Git repositories. I even had a name: GitHub." Fast forward more than 16 years later, GitHub is how people build software. With a community of more than 100 million people, developers can discover, use, and contribute to 370 million projects using a powerful collaborative development workflow[23].

2009 - USAA, a privately held bank and insurance company, becomes the first company to introduce the mobile check deposit feature, which requires a customer to photograph both sides of the check with the phone's camera. With just one branch, in San Antonio, and customers deployed all over the world, the company aggressively developed an anytime, anywhere banking strategy[24].

2009 - Al Gore makes his infamous claim in a CNN interview, "During my service in the United States Congress, I took the initiative in creating the Internet[25]."

2009 - Web 2.0 hits its stride. Google offers browser-based enterprise applications, through services such as Google Apps[26].

2011 - The Oxford Dictionary added the common texting acronym "LOL" (laugh out loud) to its official listings.

---

23. http://tom.preston-werner.com/2008/10/18/how-i-turned-down-300k.html
24. http://www.nytimes.com/2009/08/10/technology/10check.html
25. https://www.washingtonpost.com/blogs/fact-checker/wp/2013/11/04/a-cautionary-tale-for-politicians-al-gore-and-the-invention-of-the-internet
26. http://www.computerweekly.com/feature/A-history-of-cloud-computing

2013 - Out of the world's estimated 7 billion people, 6 billion have access to mobile phones. Only 4.5 billion have access to working toilets[27].

2014 - Intel and its partners announce that they will begin commercial production of its 800 gigabit per second optical network cables, based on technology developed in its Silicon Photonics lab[28].

2015 - Amazon surpasses Walmart as the most valuable retailer in the United States by market capitalization.

2015 - The average person unlocks his or her smartphone 110 times each day[29].

2015 - OpenAI was founded in December as a research organization with the aim of promoting and developing friendly AI in a way that benefits humanity as a whole. The organization has since been involved in various AI research and development projects, becoming a prominent name in the field of artificial intelligence with Chat GPT.

2016 - Driverless cars shifted into high gear in 2016. Tesla pledged to include an enhanced, completely self-driving Autopilot in all models, and Uber was the first ride-sharing company to put driverless cars on the road, starting in Pittsburgh.[30]

2017 - Facebook admits that its platform was used by Russia to sway the 2016 American presidential election and the 2017 French election.

---

27. http://newsfeed.time.com/2013/03/25/more-people-have-cell-phones-than-toilets-u-n-study-shows/
28. http://www.wired.com/2014/03/intel-mxc/
29. http://www.knowyourmobile.com/nokia/nokia-3310/19848/history-mobile-phones-1973-2008-handsets-made-it-all-happen
30. https://mashable.com/article/biggest-tech-stories-2016

2017 - DeepMind announced it was developing an AI capable of imagination. The goal being the ability to predict the outcome of a decision before making it.[31]

2018 - New features like the ECG app, irregular heart rhythm notifications, and fall detection challenge the Apple Watch 4's utility as a wrist-worn computer and steer it closer towards being a powerful medical device.[32]

2018 - 5G became the fifth generation of wireless technology to hit our mobile devices. Consumers had been experiencing maximum speeds of around 16mbps on the 4G network for the better part of a decade. 5G technology promises to deliver speeds more than 1gbps - 64 times as quick as our current connection.[33]

2019 - Social robots are filling an ever-expanding variety of roles. A 47-inch humanoid called Pepper (from SoftBank Robotics), for instance, recognizes faces and basic human emotions and engages in conversations via a touch screen in its "chest." About 15,000 Peppers worldwide perform such services as hotel check-ins, airport customer service, shopping assistance, and fast-food checkout.[34]

2020 - AI has a problem. In the quest to build more powerful algorithms, researchers are using ever greater amounts of data and computing power and relying on centralized cloud services. This not only generates alarming amounts of carbon emissions but also limits the speed and privacy of AI applications. But a countertrend of tiny AI is changing that. Tech giants and academic researchers are working on new algorithms to shrink existing deep-learning models without losing their capabilities. Meanwhile, an emerging

31. https://www.wired.co.uk/article/these-are-the-highs-and-lows-of-2017s
32. https://mashable.com/feature/best-tech-2018
33. https://www.smsglobal.com/blog/9-tech-breakthroughs-2018/
34. https://www.scientificamerican.com/article/top-10-emerging-technologies-of-2019/

generation of specialized AI chips promises to pack more computational power into tighter physical spaces, and train and run AI on far less energy.[35]

2021 - Google's DeepMind subsidiary announced it had solved a "grand challenge" in biology known as the "protein folding problem." Using its cutting-edge AI tool AlphaFold, Deepmind released the structures of 350,000 proteins. And noted that the tech will eventually be able to help identify and cure diseases.[36]

2022 - The National Ignition Facility published a paper outlining their successful attempts at creating "self-heating burning plasma." It's hoped that this breakthrough will eventually lead to commercially viable fusion reactors that will provide us with a nearly limitless, cheap, and clean source of energy.[37]

2023 - ChatGPT, a variant of the GPT (Generative Pre-trained Transformer) models developed by OpenAI, was first made available to the public in November 2022. This version of ChatGPT was based on the GPT-3.5 model and incorporated feedback and learnings from earlier versions. ChatGPT quickly garnered attention for its conversational capabilities, allowing users to interact with an AI in a more dynamic and natural manner. The accessibility of ChatGPT marked a significant milestone in making advanced AI conversational systems available to the broader public.

2023 - AI usage is widespread and varied. In terms of voice assistants, 97 percent of mobile users utilize AI-powered tools, and more than 4 billion devices currently operate on these systems. A whopping 63 percent of companies surveyed plan to increase or maintain AI and machine learning spending in 2023. Moreover,

35. https://www.technologyreview.com/10-breakthrough-technologies/2020/#tiny-ai
36. https://nerdist.com/article/best-science-tech-breakthroughs-2021/
37. https://www.forbes.com/sites/bernardmarr/2022/12/19/the-biggest-tech-innovations-and-breakthroughs-of-2022

AI technology is expected to create 12 million more jobs than it is expected to replace, and 97 million specialists are anticipated to be needed in the AI industry by 2025.[38]

2024 – In February 2024, Google introduced AI project Gemini 1.5 (formerly Bard), providing significant improvements in performance, and understanding of longer contexts. However, shares for Alphabet Inc., which owns Google, fell by up to 4.4 percent to $137.57 in late February 2024. This dip was equivalent to losing more than $70 billion in value. The significant drop in value followed a pause in Gemini's image generation after the AI appeared to refuse to create images of White people, even in historical settings such as the Founding Fathers. Not every advancement in technology works the first time. But, by the printing of this book, corrections will be made, and the probability is that Google's AI Gemini project will continue to advance Google's domination in tech.

TODAY'S CONSUMER

When these innovations from the last 60 years are listed out across several pages, two facts come into sharp focus. First, the rate of innovation is ever increasing. What would seem like a giant leap 60 years ago appears more like a small step today. Second, these innovations are changing the average consumer. Going from a world where connected technology is mostly a dream to such ubiquity where more humans have mobile phones than access to working toilets is nothing short of impressive.

We shared our 71 events with ChatGPT and asked it to write a song like Billy Joel's *We Didn't Start the Fire.*

Here is the full song:

38. https://www.mailbutler.io/blog/email/chatgpt-ai-statistics-trends/

In '61, Kilby and Noyce, their genius unchained,
The silicon chip birthed, technology gained.
No more vacuum tubes, their size and their drain,
A revolution launched, a computer's domain.
Roger Easton's vision, GPS took flight,
Tracking satellites, guiding us right.
Goodnight's SAS, statistics in sight,
From farm fields to boardrooms, shining so bright.

(Chorus) We didn't start the fire, but we lit the flame,
Innovation's our anthem, the world's never the same.
From the moon to the web, we've played the game,
The journey continues, our legacy to claim.

'69, Armstrong's moonwalk, a giant leap made,
ARPANET's birth, Licklider's crusade.
Moore's Law declared, computing cascade,
Transistors and chips, our progress displayed.
Xerox Alto's dawn, a desktop dream,
Martin Cooper's call, a mobile scheme.
Personal computers, a tech regime,
Digital photos, a Kodak theme.

(Chorus) We didn't start the fire, but we lit the flame,
Innovation's our anthem, the world's never the same.
From the moon to the web, we've played the game,
The journey continues, our legacy to claim.

(Bridge) From eBay's auctions to Wikipedia's might,
Driverless cars, AI's insight.
Google's searches, YouTube's delight,
Amazon's reign, the future in sight.

In '21, DeepMind's grand feat,
Protein folding solved, diseases to beat.

In '22, fusion's heat,
National Ignition's promise, energy replete.
ChatGPT's arrival, conversation's new height,
AI's widespread, in day and night.
Voice assistants aid, with insight,
A future powered, by AI's light.

(Chorus) We didn't start the fire, but we lit the flame,
Innovation's our anthem, the world's never the same.
From the moon to the web, we've played the game,
The journey continues, our legacy to claim.

(Outro) From silicon dreams to AI's embrace,
We've journeyed far, in time and space.
The fire still burns, in every place,
Innovation's anthem, humanity's grace.

As you consider all these amazing advances over the last 60+ years, there's no doubt that consumer expectations have changed. Technology has forced them to change with each innovation being faster, smaller, and more accessible than the last. The same expectations apply to your business.

Your customers expect faster response times, shorter wait times, more value, and less cost. The companies who figure out how to leverage AI and serve this modern consumer will thrive. And the others? Their days are numbered.

Blue Goldfish 2.0 highlights those companies doing the former, using AI to improve customer experience. Technology brought us these increasing expectations and it's the only thing that can save our businesses from them.

CHAPTER 2

WHY BLUE AND WHY A GOLDFISH?

"Technology made large populations possible, large populations now make technology indispensable."

– Joseph Wood Krutch

Blue Goldfish 2.0 brings together two symbolic metaphors to describe the intersection of technology and customer experience. To fully explain the connection, we must rewind briefly to 10th century Denmark to explain the meaning behind Blue.

King Harald Gormsson ruled Denmark in the 10th century. The medieval king was notorious for uniting all of Scandinavia and converting the Danes to Christianity. Legend has it that Harald had a nasty dead tooth that turned a darkish color over time, earning him the nickname Blåtand.

Fast forward 10 centuries to 1996. A consortium of companies including Intel, Ericsson, IBM, and Nokia came together to create a new short-range wireless standard. The companies were all working on their own solutions and wanted to avoid the debacle that was VHS versus Beta. The challenge was what to name the new universal standard.

The first step towards solving the problem came about at a conference in Toronto. Two of the engineers working on the project ended up going out for a night of drinking. Intel's Jim Kardach met up with Ericsson engineer Sven Mattisson. Kardach had been working on a program called Business-RF and Mattisson had developed a comparable standard called MC Links. Over the course of the evening, the discussion turned to history. Mattisson had just read a book called *The Longships* by Frans G. Bengtsson. The book chronicled the travels of warriors serving King Harald Gormsson. Upon learning about his nickname, Kardach perked up, "It occurred to me that this would make a good codename for the program." Blåtand in English is Bluetooth. Kardach went on to pitch the name to others in the group[39].

39. http://www.eetimes.com/document.asp?doc_id=1269737&page_number=1

After much debate and no consensus, it was decided that Bluetooth would be used a placeholder. Later, the group abandoned Bluetooth when it came time to finalize the name. It was agreed that the standard would use IBM's idea of PAN. PAN was an acronym for Personal Area Networking. But it turned out that PAN presented some intellectual property challenges. A search for PAN online delivered thousands of alternative results. Again, Bluetooth became the working name until marketing decided on a different name. It never happened. Bluetooth became the standard bearer for the new technology.

King Harald would have been proud that his 10th century nickname would become a 21st century fixture. His mark also appears on the Bluetooth logo, which is the Nordic letters H and B converge into a bind rune. Drop the mark on a blue background and you have the familiar Bluetooth logo seem on millions of devices around the world. All because of a good king and his bad tooth[40].

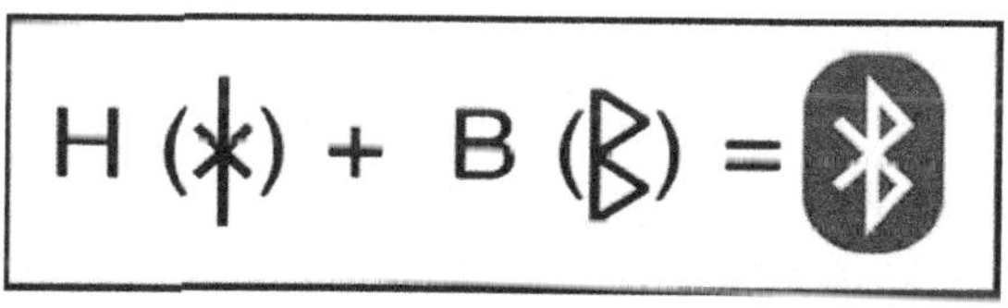

Blue became the fourth color in the Goldfish Series. It followed Purple, Green, and Golden. It is a natural name for this book because it highlights convergence, just as Bluetooth was the result of a consortium and King Harald uniting all of Scandinavia. Our convergence is the idea of big data and little data coming together with AI to deliver high-level trends and personalized insights.

40. http://factually.gizmodo.com/bluetooth-is-named-after-a-medieval-king-who-may-have-h-1671450657

NOW, WHY A GOLDFISH?

The origin of the goldfish dates to 2009. It has become a signature part of this book series. First, a goldfish represents something small, but despite its size, something with the ability to make a big difference.

The inspiration for the goldfish came from Kimpton Hotels. Introduced at the Hotel Monaco in 2001, the boutique hotel chain offers lonely travelers the ability to adopt a temporary travel companion for their stay—a goldfish.

They call the program Guppy Love. Now a signature element of the Kimpton experience, the program has gained national attention.

"The 'Guppy Love' program is a fun extension of our pet-friendly nature as well as our emphasis on indulging the senses to heighten the travel experience," says Steve Pinetti, Senior Vice President of Sales & Marketing for Kimpton Hotels and Restaurants, of which Hotel Monaco is a premier collection. "Everything about Hotel Monaco appeals directly to the senses, and 'Guppy Love' offers one more unique way to relax, indulge and promote health of mind, body, and spirit in our home-away-from-home atmosphere."

Guppy Love inspired the start of the Purple Goldfish Project. Three years later in 2012, the book *Purple Goldfish* was published. The Amazon Best-Seller was followed by *Green Goldfish* in 2013 and *Golden Goldfish* in 2014. *Blue Goldfish* became the fourth color in the series in 2016. Blue was followed by *Red Goldfish, Pink Goldfish, Yellow Goldfish, Gray Goldfish, Diamond Goldfish, Silver Goldfish,* and *Black Goldfish.*

The Goldfish is a metaphor for growth. If you study goldfish closely, they hold five lessons for business leaders today.

NOT ALL GOLDFISH ARE CREATED EQUALLY

Like businesses, not all goldfish are created equally. The average common goldfish is around three inches in length (8 to 10 cm), yet the largest in the world is almost six times that size! Imagine walking down the street and bumping into someone more than 30 feet tall (10 meters). How can there be such a disparity between your garden variety goldfish and their monster cousins? It turns out that the growth of the goldfish is determined by five factors. Part of our obsession is our firm belief that the growth of a product or service is like that of the growth of a goldfish.

The growth of a goldfish (or your product or service) is affected by the same five factors:

#1. SIZE OF THE ENVIRONMENT = THE MARKET

GROWTH FACTOR: The size of the bowl or pond.

RULE OF THUMB: Direct correlation. The larger the bowl or pond, the larger the goldfish can grow. The smaller the market, the lesser the growth.

#2. NUMBER OF GOLDFISH = COMPETITION

GROWTH FACTOR: The number of goldfish in the same bowl or pond.

RULE OF THUMB: Inverse correlation. The more goldfish, the less growth. The less competition, the more growth opportunity.

#3. THE QUALITY OF THE WATER = THE ECONOMY

GROWTH FACTOR: The clarity and amount of nutrients in the water.

RULE OF THUMB: Direct correlation. The better the quality, the larger the growth. The weaker the economy or capital markets, the more difficult it is to grow.

FACT

A malnourished goldfish in a crowded, cloudy environment may only grow to two inches (five centimeters).

#4. THEIR FIRST 120 DAYS OF LIFE = STARTUP PHASE OR PRODUCT LAUNCH

GROWTH FACTOR: The nourishment and treatment they receive as a fry (baby goldfish).

RULE OF THUMB: Direct correlation. The lower the quality of the food, water, and treatment, the more the goldfish will be stunted for future growth. The stronger the leadership and capital as a start-up, the better the growth.

#5. GENETIC MAKEUP = DIFFERENTIATION

GROWTH FACTOR: The genetic makeup of the goldfish.

RULE OF THUMB: Direct correlation. The poorer the genes or the less differentiated, the less the goldfish can grow. The more differentiated the product or service from the competition, the better the chance for growth.

FACT

The current Guinness Book of World Records holder for the largest goldfish hairs from The Netherlands at a lengthy 19 inches (50 centimeters). To put that in perspective, that's about the size of the average domestic cat.

Which of the five factors can you control?

Let's assume you have an existing product or service and have been in business for more than six months. Do you have any control over the market, your competition, or the economy? NO, NO and NO. The only thing you have control over is your business' genetic makeup or how you differentiate your product or service. In goldfish terms, how do you stand out in a sea of sameness.

In the future, where AI transcends its role as a mere tool to become a strategic ally, it magnifies the growth factors of the business ecosystem, akin to how a goldfish flourishes in an ideal environment. This symbiosis promises a realm where businesses can expand beyond traditional market boundaries, mirroring a goldfish's growth in an expansive, nourishing habitat. AI-driven differentiation, like the unique genetics of a thriving goldfish, enables businesses to carve out distinctive niches, fostering unprecedented growth and innovation. As we navigate this future, the fusion of AI with fundamental growth principles heralds a new era of business, unlocking potentials as limitless as the growth of a well-cared-for goldfish in a boundless pond.

CHAPTER 3

A LITTLE SOMETHING EXTRA

"We picked up one excellent word - a word worth traveling to New Orleans to get; a nice limber, expressive, handy word—lagniappe"

- Mark Twain

An important word to know as you read the Goldfish Series is lagniappe. Lagniappe is a Creole word meaning *"the gift"* or *"to give more."* The practice originated in Louisiana in the 1840s whereby a merchant would give a customer a little something extra at the time of purchase. It is a signature personal touch by the business that creates goodwill and promotes word of mouth.

According to *Webster's*:

LAGNIAPPE (lan'yəəp, lăn-yăp') *Chiefly Southern Louisiana & Mississippi*

1. A small gift presented by a store owner to a customer with the customer's purchase.
2. An extra or unexpected gift or benefit. Also called *boot.*

Etymology: Creole < French la, the + Spanish ñapa, lagniappe < Quechuan yapa. Interesting fact- Napa comes from yapa, which means "additional gift" in the South American Indian language, Quechua, from the verb yapay "to give more."

ENTER SAMUEL LANGHORNE CLEMENS

Mark Twain was so smitten with the word that he wrote about it in his autobiography, *Life on the Mississippi*:

> We picked up one excellent word–a word worth traveling to New Orleans to get; a nice limber, expressive, handy word – 'lagniappe.' They pronounce it lanny-yap. It is Spanish–so they said. We discovered it at the head of a column of odds and ends in the [Times] Picayune [newspaper] the first day; heard twenty people use it the second; inquired what it meant the third; adopted it and got facility in swinging it the fourth. It

> has a restricted meaning, but I think the people spread it out a little when they choose. It is the equivalent of the thirteenth roll in a baker's dozen. It is something thrown in, gratis, for good measure. The custom originated in the Spanish quarter of the city. When a child or a servant buys something in a shop–or even the mayor or the governor, for aught I know–he finishes the operation by saying– 'Give me something for lagniappe.' The shopman always responds; gives the child a bit of licorice-root, gives the servant a cheap cigar or a spool of thread, gives the governor–I don't know what he gives the governor; support, likely.

ACTS OF KINDNESS

Another way to think of lagniappe is as an act of kindness. There are three types:

1. **Random Act of Kindness** - we've all seen this before. Good deeds or unexpected acts such as paying tolls, filling parking meters or buying gas for consumers. They are usually one-off, feel-good activations. A random act of kindness draws upon gift economy principles. Giving with no expectation of immediate return, except maybe for potential PR value.
2. **Branded Act of Kindness** – next level 2.0. Here the item given is usually tied closely with the brand and its positioning. It's less random, more planned, and potentially a series of activations. This has the feel of a traditional marketing campaign.
3. **Lagniappe Act of Kindness** – 3.0 stuff. Kindness imbedded into your brand. Giving little unexpected extras as part of your product or service. This is rooted in the idea of 'added value' to the transaction. Not a one off or a campaign, but an

> everyday practice that's focused on customers of your brand. The beauty of creating a Goldfish as a branded act of kindness is that there is no waste. You are giving that little extra to your current customers.

As you dig into the three types of Blue Goldfish over the next three sections, you'll see how companies who do the little extras–the lagniappe–see impressive results.

PART II:

THE FIRST R

CHAPTER 4

RELATIONSHIP

"Computers are like bikinis.
They save people a lot of guesswork."

– Sam Ewing

The year was 1981—the same year the first Space Shuttle rocketed from the Florida coast. A few weeks after STS-1 launched in April a sixteen-year-old named Michael, then a high school student, took a summer job selling newspaper subscriptions for *The Houston Post.*

Michael recounted his time at the paper, "At the time, the newspaper gave its salespeople a list of new phone numbers issued by the telephone company and told us to cold call them. It struck me as a pretty random way of approaching new business."[41]

Michael soon noticed two common threads. New subscribers tended to be people who had just moved into new homes or who had just married. Armed with this realization, Michael hired two high school friends to identify how he could find people getting mortgages or getting married.

Armed with these qualified leads, Michael penned a personalized letter to these prospective subscribers. He would sell thousands of subscriptions and earn more than $18,000 that summer. To put that into perspective, $18K was more than his high school economics teacher made the entire year.

Why did this work so well? Michael uncovered the core ingredient of customer segmentation, which is knowing your customer. By knowing more about his ideal customer, Michael was able to eliminate the guesswork and, knowing his prospect would likely convert, spend more time connecting with them on a personal level.

By the way, Michael would later head off to college at the University of Texas. In his sophomore year, he would start a computer

41. *Direct from Dell: Strategies that Revolutionized an Industry,* HarperCollins 1999

company focused on customized personal computers (PC's). That company was called... Dell Computers.

This lesson that Michael Dell learned is even truer today than it was in 1981. Technology has paved the way for every company to know more about their customers and build deeper relationships with them. Now it's your turn to leverage it.

CHAPTER 5

PERSONALIZATION

"Technology is nothing. What's important is that you have a faith in people, that they're basically good and smart, and if you give them tools, they'll do wonderful things with them."

– Steve Jobs

Personalization is a popular term in business today and before we go any further, we'd like to point out the irony. Business was always personal. In the world where individual artisans or merchants owned their shops, they knew each customer. Recommendations and personal touches were commonplace.

Today we use the word personalization to describe our desire to go back to doing business on a personal level, but often forget it's not about segments or cohorts. To personalize is to design or produce something to meet someone's individual requirements. The key is the individual.

In our current landscape focusing on the individual can be almost impossible for companies. Stymied by legacy technology and distributed workforces, personalization is difficult. Yet the companies who overtake those barriers tend to see impressive results.

Connecting the Dots

Personalization is about connecting the dots. Specifically, it's about bridging the gap between online history and offline behavior. Retailer Tory Burch has developed a system for customers and employees that marries both. Here is a story about their system told by Doug Logue:

> Let me paint you a picture. I wake up in Toronto, on a bachelor party and have this nagging notion that something is wrong...Then it hits me. I forgot to order my pregnant wife's gift before leaving the country just days before her birthday. Panic immediately sets in. Now keep in mind I didn't bring a laptop or tablet and chose not to get an international data plan for my phone.
>
> I race down to the hotel lobby and jump on one of their ancient machines running Windows Vista with

Internet Explorer as the only browser choice—my nightmare begins. I pull up Gmail, run a search for 'bday ideas,' and click on the link for a pair of Tory Burch shoes. Find the right size, add to cart, choose expedited shipping, proceed to checkout... and realize that I forgot to grab my wallet when I ran out of my room.

No sweat, PayPal to the rescue! Or so I thought. After checking out with PayPal, I receive a confirmation email from Tory Burch where I notice the shoes were shipped ground and won't arrive until after my wife's birthday. By this time my friends have started to gather in the lobby and we're ready to head to the Blue Jays game. I quickly forwarded the email to customer service and asked them to please change the shipping option.

After the game and a few pints, we end up at the Loose Moose (it's a bar, I swear) where they have free Wi-Fi. I check my email and read the response from customer service. *'Unfortunately, Mr. Logue, when you checkout with PayPal you don't have the option to upgrade your shipping method.'* I reply asking them to cancel my order, so that I can repurchase the shoes using my debit card and choose two-day shipping. *"Unfortunately, Mr. Logue, we can't cancel nor modify your order at this time."*

Frustrated, I decide that it will be easier to just wait until I get home to repurchase the shoes at an actual Tory Burch store.

Two days later, and with just 24 hours left, I make my way to the store and immediately have a much better experience than I'd been having online. A very nice woman named Michelle introduces herself and asks if she can assist me with anything. I begin to explain my situation and ask what she can do. Michelle tells me that once my online order arrives to simply bring it into her store and she'll return them no problem. She grabs a new pair of shoes, swipes my debit card, and tells me that it isn't going through… Wonderful.

She remembers me saying that I just returned from Canada and thinks my bank may have put a hold on my card. I give my bank a call, get put on hold, transferred, hold again, and then Michelle emerges from the back of the store like some kind of customer service angel… *'Would you like this Corona or perhaps a glass of wine while you get this sorted?'*

What an unexpected surprise! My mood immediately changed. Not because of the free booze, but because Michelle saw that I was having a rough day and went above and beyond to make me happy. Eventually, Chase lifts the hold on my card, Michelle charges it and wraps my wife's gift. Now this is where most transactions end, but she was just getting started.

The next day, I receive an email from Michelle thanking me for my purchase and reminding me that whenever the original order shows up, I can bring it in and she'll return it for me. A few days later my wife and I head back to Tory Burch. We walk in and Michelle greets me by name and introduces herself to my wife. While she's returning the original order, my wife begins trying on shoes and I take a seat. Within seconds, I have a Corona in one hand and an iPad in the other. These people are good. Michelle knows that if she can make me comfortable and keep me preoccupied, my wife will have more time to drain the bank account.

After the return was complete and my wife was done browsing, we left the store. A few days later, I receive another email from Michelle. *'Hi Doug, it was a pleasure meeting your wife, Lindsay. Just wanted to check in and make sure she was completely satisfied with her new shoes. Also, I noticed her trying on the navy-blue Kaitlin Wedges in 8.5. Would you like me to put these on hold for a future purchase?'*

This was the first time I've seen online marketing being fused with an offline transaction. Marketers like myself, use software to show you ads for products you recently searched for or viewed in hopes of getting you to complete your purchase. The fact that Michelle was doing this manually really impressed me.

One of the things I liked most was the personalization of this email. Michelle used my name, my wife's name, the type of shoes she tried on, the size of the shoe, the color, and included a picture. I'm sure that Tory Burch has some type of CRM that populates the email with my basic information, but the shoes my wife tried on didn't have a tracking pixel embedded in them. That information had to be entered manually. I'm not sure if this is common practice at their stores or exclusive to Michelle. Either way, I highly recommend other retailers find a way to merge offline interactions with online communication and vice versa.

These tactics are usually only seen in the B2B space, where the sales cycle is much longer than an in-store retail purchase. However, I believe these practices will become commonplace in the very near future... Merging the offline and online worlds is something I'm extremely interested in and will be exploring in the coming months.

Needless to say, I follow up with Michelle and thank her for offering to hold the shoes. At the time, I wasn't ready to make a new purchase. About a month later, I receive a phone call from Michelle. *'Hi Doug, this is Michelle from Tory Burch. Is now a good time to talk?'* Normally, I would identify this as a sales call and say no. However, my previous experience with Michelle kept me on the line. *'I just wanted to call and let you know the shoes your wife was interested in are now 30 percent off.'*

Did a sales rep really call me and tell me a product was on sale? Sure, I get email coupons and alerts for generic sales all day long, but once again this was personal and specific. Since the holidays were right around the corner, I told Michelle to move forward with the purchase. Michelle saved me from having to fight holiday crowds and traffic. She saved me from having to distinguish what "super cute" shoes my wife had been trying on that day. And she saved me 30 percent! This was a new level of customer service that I wasn't familiar with but could definitely get used to. This experience has coined a few sayings in the Logue household...

'Honey, I want... just email the link to Michelle.'
'This would never happen at Tory Burch.'
'$300 Coronas just taste better.'

Soon stores and physical items will send reminders to your mobile device about which pair of shoes your wife tried on. Until then, find a Michelle[42].

42. https://medium.com/@douglogue/tory-burch-sells-300-coronas-3944d1c21a56#.sa7lg8tm4

The Tory Burch system is called Client Book. The tablet-based system aggregates their data to provide detailed customer profiles that include personal preferences, online activity, and previous order history. Store associates such as Michelle can find sizes, check availability, view recommended items, and add products to a shopping cart. Associates can communicate with customers via e-mail on their tablet. This allows associates to thank clients for stopping by, notify customers about sales, recommend products and schedule appointments.

According to Matt Marcotte, Head of Global Direct to Consumer for Tory Burch, "Client Book enables customers to keep track of previous orders, wish lists and other information from online accounts. A shopper might put something into her wish list late at night, for instance, and a store associate keeping track could have that item ready on the shopper's next store visit or prepare recommendations for other products."

Client Book also prompts associates to send a handwritten card in advance of a customer's birthday. Marcotte shared the following story at Shop.org's Annual Summit:

> In one case, a married man came into one of our stores, asking for assistance from a store associate. He wanted to purchase a birthday gift for his wife. He explained that he didn't know where to turn but saw the company's logo on his wife's shoes and assumed that she shopped at the store. The associate looked up his wife's name in Client Book; turns out she's shopped at Tory Burch both in person and online. Anything she bought in the last two years showed up, with photographs. So, the associate made suggestions and he goes off happy. The wife comes in the next day and asks to see sales associate. That's never good—usually it means that something's gone wrong. The woman

> says, 'I don't know what you did, but it's the first time in 20 years he's ever gotten in right.' So, the associate shows her Client Book. No return necessary; the husband is a rock star. She leaves. Three hours later, her friend comes in. Turns out the wife was talking about Client Book at lunch and she just had to know what it was all about[43].

Tory Burch's OPEN strategy—On-demand, Personal, Engaged, and Networked—is revolutionizing retail engagement, and AI stands as a pivotal ally in this transformation. Imagine AI systems intricately designed to mimic human interaction, offering personalized shopping experiences on-demand. These systems could analyze customer preferences, past purchases, and browsing habits to provide tailored recommendations, much like a personal shopper. Engaging customers through AI-driven Client Books could elevate the user experience to new heights, ensuring each interaction feels unique and genuine.

Moreover, AI can foster a networked environment by connecting users with similar tastes or recommending complimentary products, thereby enriching the community aspect of shopping. From an AI perspective, blending human-like empathy with data-driven precision could be the next frontier in retail, offering customers a seamless, engaging, and deeply personalized shopping journey.

That's the power of personalization – both human and AI style.

Heroic tales of customer service like Doug Logue's are often romanticized because they aren't common practice, even if they seem like common sense. Companies all over the world are waking up to this trend, taking the uncommon path and using technology to deliver personalized service to customers.

---

43. http://www.zdnet.com/article/at-tory-burch-a-clear-eyed-focus-on-the-customer/

Next Best Action

Much like the Westpac example from earlier in this book, Erste Bank, headquartered in: Vienna, Austria uses analytics to determine the next best action for customers. Erste uses the SAS Customer Intelligence suite to analyze existing data and to recommend next best actions to customers via outbound and inbound touchpoints.

SAS had this to say about next best offers in a recent whitepaper:

> Next best action is not a new concept. It is a fundamental marketing approach to understanding customers and using the knowledge to anticipate their needs. Businesses and merchants have been successful with this approach for years; especially more mature businesses. What has changed is the reason for conducting such marketing, the ability to market more effectively with a large number of customers and the ability to measure its effectiveness.
>
> In the past, understanding each customer's needs required a person to manually track information about the customer. When the number of customers is small – as with the traditional corner store, an accountant who prepares taxes for her clients or a local barber – this can be done fairly easily just by communicating and working with each customer. While this approach was also possible for a larger number of customers, it was often limited to businesses that could absorb the high cost and resource requirements, such as high-end hotels and resorts.
>
> Technology advancement is driving the ability to implement a next best action strategy with a large number of customers without significantly increasing headcount or spending inordinate amounts of

> money on each customer. With the prevalent use of point of sale (POS), call center systems, enterprise systems and Internet applications, organizations are able to collect more data about customers than ever before. With the low cost of data storage, most organizations can keep and maintain this data to provide a rich history for each customer. With point of contact systems, organizations can identify customers while the interaction is taking place and collect up-to-date information to augment existing data. Companies can now treat each customer as an individual, while actually improving productivity by quickly using all relevant data.

Next best action analytics isn't just a theoretical approach, rather it's driving impressive results for Erste Bank. The bank makes these personalized offers through its call centers and online channels. They report a 95 percent success rate in determining the next best product for a customer using their data-driven approach. Getting this offer right is moving the needle for Erste as purchases for the next best product are up to 16 times more than the average sale ratio.

SAS is changing the game in customer engagement using AI. Imagine you're shopping online, and every suggestion feels like it's handpicked just for you. That's AI in action, helping businesses understand what their customers love, creating personalized marketing campaigns that hit the mark every time. Take a fashion retailer, for instance – they can tailor their email campaigns so well that customers feel like they're getting recommendations from a friend, not a computer. It's not just about making things faster; it's about making shopping feel personal and genuinely enjoyable. That's the magic with AI – turning shopping into a personalized experience for everyone.

The integration of Next Best Action (NBA) strategies using AI represents a transformative leap in how businesses engage with and understand their customers. By harnessing AI, companies can sift through vast amounts of data in real-time, deriving nuanced insights into customer behavior, preferences, and potential future actions. This allows for the delivery of personalized recommendations and services that are not only aligned with the customer's current needs but are also anticipatory in nature, enhancing the customer experience and fostering loyalty. AI-driven NBA systems are dynamic, continuously learning from new data to refine and optimize suggestions, ensuring that businesses remain agile and responsive in their customer interactions. This integration marks a pivotal shift towards more intelligent, data-driven decision-making, enabling businesses to achieve a competitive edge by offering unmatched personalization and relevance in their customer engagement strategies.

Art and Science

Looking to companies like Westpac and Erste Bank you might conclude that personalization is mostly a science. Companies today are also looking to subjective data as a part of their personalization efforts. One such company is Stitch Fix, which offers online personal stylists for women.

Stitch Fix has a subscription-based service that sends a monthly selection of five items to each customer's door. A combination of sophisticated algorithms and personal touches from stylists, helps Stitch Fix create a monthly delivery to meet each client's budget, taste, and lifestyle. Once each month's "fix" arrives, customers keep what they like and send back what they don't. There's no obligation to keep any or all items, although there is a discount should a customer choose to keep all five items. The return process offers a way for Stitch Fix to learn what customers like and take this into account for future shipments.

Stitch Fix brings together an unusual blend of art and science to deliver one-on-one personalization at scale[44].

When asked about its personalization efforts, Stitch Fix COO Julie Bornstein told NRF:

> Personalization and big data are buzzwords that everyone talks about, but it's very hard to operationalize these concepts. At Stitch Fix we are able to deliver a "fix" of five items that have been selected to fit each individual client's budget, taste, and lifestyle. It is personalization at a whole different level, and it's very exciting. In some ways, it's the combination of big and small data that is enabling us.

The combination of big data and small data is at the heart of personalization. Many companies are great at identifying big-data trends across their customer base. Magic happens, however, when you use those trends to see little-data behaviors in one customer as an indicator and respond accordingly.

Stitch Fix is redefining its use of AI with an "expert-in-the-loop" approach. This innovative strategy blends the best of both worlds: the efficiency and scalability of AI with the creativity and insight of human expertise. Essentially, AI generates initial content like ad headlines or product descriptions, which are then refined by human professionals to ensure they perfectly capture the brand's voice and meet customer expectations. This collaboration not only saves time but also enhances the quality of the output, making the shopping experience more personalized and engaging. It's a smart blend of AI's speed with a human touch, ensuring that Stitch Fix stays ahead in offering a unique and satisfying customer experience.

44. http://www.forbes.com/sites/deniselyohn/2015/12/02/stitch-fix-combines-high-tech-and-high-touch-to-transform-retail/

Verizon "Rep Guidance" Helps Offer Personalized Service

Verizon personalizes its service through a desktop solution called Rep Guidance. The system shows reps detailed customer information such as how much data they use, what shows they watch and which TV equipment they use the most[45].

Rep Guidance allows call center representatives to foster more intelligent, better-connected conversations with customers on one single screen. Over 100 Verizon customer representatives worked together with internal Customer Experience and Technology teams to build the solution.

"We're trying to use data to be predictive and anticipate what the problem is. We realized that when a customer is calling in, they assume we know all the details about their situation." says Miguel Quiroga, executive director for Verizon Telecom[46].

How does it work? For example, a customer calls in to talk about their account. They inquire about upgrading their TV package or inquire about available speed upgrades. Rep Guidance provides meaningful insight about the customer, allowing the representative to right size their TV viewing and broadband needs.

According to Tami Erwin, Verizon President of National Operations, "This cuts down on those moments where one family member would call to switch their video package, only to find out later that their family member or loved one will lose some of their favorite channels. Or a customer calls to save money on their bill, our reps now have true insight into their preferences to provide guidance."

---

45. http://www.forbes.com/sites/micahsolomon/2015/02/18/verizon-yes-verizons-massive-customer-service-initiative-the-inside-scoop/
46. http://arstechnica.com/business/2015/05/verizon-fios-reps-know-what-tv-channels-you-watch/

Verizon's integration of AI chatbots into its customer service operations represents a significant shift in how the company manages customer interactions and overall service quality. The use of AI in their call centers and customer service processes has been multifaceted and innovative.

Globally, chatbots were forecasted to save businesses 2.5 billion hours of work in 2023. They are increasingly being used for customer support, with 35 percent of users preferring chatbots for resolving complaints or issues. Additionally, 62 percent of consumers prefer connecting with a bot rather than waiting in line to connect with a human agent. These statistics indicate a significant shift towards AI and chatbots in customer service. By 2027, chatbots are predicted to become a primary channel for customer service in a quarter of all businesses. In this context, Verizon's AI-driven approach to customer service, with its focus on personalized, efficient, and integrated solutions, positions the company at the forefront of this transformative trend in customer engagement and support.

Well Fargo ATM's Adapt to Your Personal Preferences

Using AI and predictive analytics, Wells Fargo's automated teller machines (ATMs) present customers with an array of touchscreen buttons all based on their previous transactions and preferences. The solution provides a layout that's faster and easier for customers to use. The ATM interface has been completely redesigned, featuring a sleek, modern design.

When a Wells Fargo customer inserts their card into the ATM and successfully enters their PIN, the customer's account balances appear automatically on-screen. On the main screen, the ATM presents two green buttons, both customized specifically for each customer based on previous transactions. If, for instance, a customer makes frequent deposits into their family checking account and opt for no receipt, that becomes one of the green buttons. If the customer always withdraws $60, that's a button too. Immediately below the two big green buttons are three more black buttons, also customized to reflect each customer's' common ATM behaviors. Putting buttons right on the homepage for what customers do most is smart UX, and it saves time.

"We redesigned our ATMs based on what customers tell us they want — quicker, easier, and more convenient," said Alicia Moore, head of Wells Fargo ATM Banking. "The ATMs really know you better. It's not just about putting colors together and putting a design together. It's about creating and continuing to create an experience.[47]"

Wells Fargo's Fargo™ Virtual Assistant, integrated into their mobile banking app, represents a significant advancement in customer service through AI technology. Utilizing Google's Dialogflow, a conversational AI, Fargo™ provides users with instant, intuitive assistance for everyday banking questions. This innovative tool exemplifies Wells Fargo's commitment to digital transformation, offering an enhanced, real-time banking experience. By leveraging the vast data from consumer interactions, Fargo™ is designed to understand and respond to customer needs effectively, demonstrating Wells Fargo's focus on harnessing AI to streamline customer interactions and improve overall banking convenience.

The AI initiatives actively in the works with Wells Fargo span from enhancing customer interaction through virtual assistants to employing sophisticated AI models for lending decisions and financial health management. The bank is also innovating in platform-based banking, underlining its commitment to leveraging AI for growth and improved customer experiences.

---

47. http://thefinancialbrand.com/27694/new-wells-fargo-atms-can-predict-what-people-want/

Pep Boys Develops a System for Personalization

Pep Boys sought to personalize their marketing. The first step involved creating a simple tracking mechanism. Through a partnership with Acxiom, Pep Boys set up a system for capturing data. The process involved deciding which data to collect and the method for organizing it. They then moved into designing the rewards program and building the system.

According to CMO Ron Stoupa, "Once we had the data, the fun started. We were able to start running tests and building different models. The key was to get into 'surround sound marketing.' It is about the voice the customer wants to hear, the vehicle they prefer, the offer they want and the day / time to deliver it. We had the mechanism to really understand the consumer and a strong team to build a relationship with them.[48]"

OKL Boosts Sales with Killer Content and Smart Product Suggestions

Top eCommerce brands are discovering that they can build followings, boost conversion, and ensure retention by providing practical personalized information.

An example of this is One Kings Lane (OKL). OKL is a furniture and home decor flash sale site. Positioning itself as a lifestyle brand, OKL has become a site worth checking primarily for content. Once a visitor is interested, the site then uses its advantageous positioning as an authority on lifestyle products to sell its wares.

According to *Retention Science,* what can other brands learn from One Kings Lane?

48. http://www.forbes.com/sites/kimberlywhitler/2015/06/28/sports-authority-c-mo-ron-stoupa-on-creating-a-data-centered-organization/2/

1. Useful (and beautiful) curated content.

   It's not enough to put up a few images of sofas and call it a decorating blog. One Kings Lane's style blog, Live Love Home, goes the extra mile by offering beautifully styled and curated content. OKL's blog is an entity unto itself, offering everything from design tips to the perfect recipe for a night in. The photos, staging, and copy are high quality, lending an upscale feel to the site.

2. Smart product suggestions.

   At the bottom of OKL's blog posts, you'll find relevant product selections from the flash sale side of the site. One post offers a tour of a well-designed upstate New York home. The post ends with product recommendations meant to emulate the decor featured throughout the post with the prompt, "Shop Their Modern Cabin Look." Additional editor's picks along the right sidebar of the site round out the product suggestions without smothering the reader in marketing.

Hertz Uses Predictive Analytics to Drive Personalized Upgrade Offers

Hertz, drives more revenue from every customer by offering personalized offers. Their customer intelligence system gets utilized when a customer interacts with a customer-care representative, checks in at a rental counter, or receives a receipt. The platform analyzes data collected through loyalty programs and responses to past offers.

According to DataInformed, "The sophisticated system uses predictive analytics to assess the probability for conversion. Then, the system calculates the overall expected revenue by combining the profit from the offer and the overall probability to convert. For example, if a customer has 10 percent chance for converting on an offer worth $50 and 90 percent chance converting on an offer worth $20, the system will choose the latter."

Based on the type of interaction, the system can identify what kind of promotion is likely to work best. For example, if a customer is qualified for "buy one get one free," he may still get another offer if he refused a similar offer in the past. Another example is that if someone is not yet a member of Hertz's loyalty program, his receipt may remind him to enroll.

Hertz's CIO, Joe Eckroth, says that his company's strategy is to give its customers "a wrapper of technology around your entire experience with Hertz."

United Airlines Delivers Personalized Offers Within 200 Milliseconds

As part of check-in at United, a passenger receives a variety of upgrade offers. In the past, the airline would use a "collect and analyze" approach. Given their analysis, the airline would use information about customers' choices about those items in an aggregated fashion. According to Scott Wilson, United's vice president of ecommerce and merchandising, "[We'd] see what the most successful products were, and market with those [insights] in mind."

United now has a more personalized focus of leveraging customer data called, "collect, detect, act." Wilson touches on the difference, "Now we look at who the customer is and his or her propensity to buy certain products." Over 150 variables are now assessed in real time to determine an individual's likelihood to act.

According to an article by *Fortune.com*, personalized offers are delivered in about 200 milliseconds. There is a dynamically generated offer that's tailored to the customer. Its terms, on-screen layout, copy, and other elements will vary based on an individual's collected data. According to Wilson, this personalized approach has resulted in a 15 percent increase in year-over-year ancillary revenue[49].

John Deere Uses AI Across the Agricultural Landscape

John Deere, a leading agricultural technology company, has been at the forefront of revolutionizing precision agriculture by integrating space technology, AI, and advanced data analytics to provide farmers with actionable insights. The company leverages satellite imagery to monitor crops, assess field conditions, and optimize resource allocation for improved yields. By pinpointing variations in soil moisture, nutrient levels, and plant health, John Deere enables farmers to make informed decisions regarding irrigation, fertilization, and crop protection, thereby minimizing waste, maximizing yields, and reducing the environmental impact of agriculture.

In addition to space technology, John Deere has made significant investments in electrification, automation, and autonomy to enhance farming efficiency and sustainability. The company has developed electric tractors powered by batteries with low maintenance costs. Furthermore, John Deere uses AI algorithms and machine learning techniques to analyze data and provide farmers with insights that can help them make better decisions. The company has also unveiled its first fully autonomous tractor, which is designed to complete tasks on time, every time, and at a high level of quality.

John Deere's efforts in developing AI solutions are part of a larger trend across the agricultural landscape. The company has been planting and growing data seeds for over two decades and has made strategic acquisitions, such as the machine learning company Blue River Technologies, to bolster its AI and deep learning capabilities. John Deere remains focused on developing and delivering AI-powered solutions that are crucial for the future of farming,

49. http://fortune.com/2014/06/19/big-data-airline-industry/

including autonomous production systems and advanced technology that provides farmers with the flexibility to manage pressing tasks within their operation.

John Deere's innovative use of AI and automation is revolutionizing agriculture, offering farmers the tools and insights they need to optimize their operations, improve yields, and contribute to a more sustainable and efficient future in farming.

Carbon Robotic Uses AI to Transform Weed Killing

Carbon Robotics has created an Autonomous Weeder, a high-tech solution that leverages artificial intelligence to identify and eliminate weeds while leaving crops unharmed.

Carbon Robotics' Autonomous Weeder uses advanced computer vision and deep learning algorithms to distinguish between crops and weeds. It operates by projecting concentrated light beams (lasers) onto weeds, effectively zapping them without the use of chemical herbicides, thereby reducing environmental impact and improving crop yield. This technology allows for precise weed control, significantly reducing labor costs and enhancing the sustainability of farming practices.

Five Additional Examples

Here are five additional examples of companies leveraging data and AI to offer personalized customer experiences, showcasing the breadth of application across various sectors:

1. **Netflix's Personalized Viewing Recommendations:** Netflix uses complex algorithms to analyze viewing patterns, search histories, and ratings to provide personalized recommendations. This tailored approach keeps users engaged by offering content that aligns with their preferences.

2. **Spotify's Discover Weekly Playlists:** Spotify leverages AI and machine learning to curate personalized playlists for users. The "Discover Weekly" feature analyzes listening habits, favorite genres, and frequently played tracks to introduce new music tailored to each user.
3. **Amazon's Personalized Shopping Experience:** Amazon utilizes AI to offer personalized shopping experiences. The AI makes recommendations based on past purchases, search queries, and item ratings. This not only improves the shopping experience by making it more relevant to the individual shopper but also increases the likelihood of purchases.
4. **Starbucks' Personalized Marketing Offers:** Starbucks uses its mobile app and rewards program to collect data on customer preferences and purchase history. Using this information, the app makes personalized offers and tailored recommendations, such as suggesting new products like those previously enjoyed.
5. **Duolingo's Adaptive Learning Paths:** Duolingo employs AI to customize learning paths for each user in real-time. By analyzing correct and incorrect answers, the app dynamically adjusts upcoming exercises to focus on areas needing improvement. This personalized approach helps users stay engaged by continuously adapting to their evolving needs.

All these examples demonstrate how personalization, powered by AI and data analytics, can significantly enhance customer experiences across different platforms and industries, driving engagement, loyalty, and business growth.

As you can see, the possibilities for enhanced customer experience are endless when you leverage AI to combine big data and little data into a personalized interaction.

CHAPTER 6

PERSONAL DATA / BEHAVIOR CHANGE

"Technology is teaching us to be human again."

– Simon Mainwaring

Some might argue technology is making us less human. Others would argue just the opposite. No matter which opinion you share, you can't dispute the rising trend of personal data collection and its effect on behavior change.

The smart people of TrendWatching.com, who came up with info-sense, coined another term in data myning. The intentional misspelling highlighted how individuals, not just companies, would have awareness and ownership of their own data.

The overarching lesson of data myning for customer experience is empowerment. We believe companies who use personal data to empower their customers have an opportunity to create transformative change for those customers. With empowerment and then success, gratitude and loyalty are often the next set of emotions.

FITBIT ENABLES PERSONAL BEHAVIOR CHANGE

Digital consultancy AnswerLab gave every employee a Fitbit tracker as a holiday gift. As a digital shop they tested the devices in advance and ultimately chose Fitbit because it embraced basic human behaviors. CEO Amy Buckner had this to say in *Inc*:

> Fitbit is so in tune with human psychology that many of us have grown a deep affection for the small clip-on gadget. One of our senior researchers caught a cab in Florida, realized her Fitbit had fallen off, made the driver return to the research facility where she originated, searched the elevator, sprinted back to the eye tracking room, found the Fitbit on the floor, and dashed back to the cab.
>
> We long for the Fitbit greeting, 'Howdy, Amy; walk me,' like coming home to a much-loved pet. In fact, many of us wonder if we exercise without the Fitbit on if we've really exercised at all. Our attachment to

> our Fitbits makes us huge promoters of the product to friends, family, and colleagues[50].

Empowering customers can't be the entirety of your customer experience strategy, however. Fitbit backs up their product's user experience with world class customer support. Browsing a Fitbit thread on the popular discussion website Reddit uncovers impressive examples of customer service.

User Hhumerus shared their story to open the thread:

> I got a fitbit flex for my birthday and less than a week later, it fell off while I was at work and was nowhere to be found (I'm about 95 percent sure it somehow ended up in the trash that had been taken out right after I lost it, it wouldn't sync up to my phone no matter where I went).
>
> Anywho, I figured it was a long shot, but I emailed fitbit about my situation and requested a new one. I included pictures of my receipt and a screenshot of the last time my fitbit synced to me phone. Within 24 hours, I had an email back from them saying that they would send me a new one free of charge!
>
> I honestly didn't have any high hopes about getting a new one from them, but I was definitely pleasantly surprised[51].

The list of positive stories continues with user "Mizscarlett" who lost hers in a Walmart store and received a free replacement within

50. http://www.inc.com/amy-buckner-chowdhry/user-experience-best-practices-how-fitbit-gets-it-right.html
51. https://www.reddit.com/r/fitbit/comments/30lpwv/fitbit_honestly_has_the_best_customer_service/

a week. User "Superman_does_good" lost theirs in an airport and received a free replacement, ending his Reddit comment with "They're awesome."

The lesson here is once you empower customers with devices and data that are integrated into their lives don't fumble the ball. Rise to the occasion with phenomenal customer support. Your loyal customers, who happen to wear your device every day, will share their positive experiences with others, just as they did on Reddit. That's better than any advertising you can buy.

The benefits of personal data aren't just restricted to fitness. John Hancock allows policyholders to turn healthy activity into lower life insurance premiums. Hancock has partnered with Vitality, a global wellness firm on a groundbreaking program.

How does it work? Members of the program receive a Fitbit monitor as one way to track their fitness. Activities allow members to accrue Vitality Points. Taking a cue to gaming as you accumulate points, your status rises. The four levels are bronze, silver, gold, and platinum. The higher your status, the more you save each year on your life insurance premiums. Vitality points also allow you earn benefits at stores like REI, Whole Foods, and hotel chains like Hyatt.

What's at stake for customers participating and sharing their personal data? The most active customers can earn up to a 15 percent reduction on their premiums.

And for those uncomfortable with sharing some data, Michael Doughty, John Hancock's president, told The Times: "You do not have to send us any data you are not comfortable with. The trade-off is you won't get points for that."

In the auto industry, Progressive offers its customers a mileage-based tracking device called Snapshot; those who opt-in can save up to 30 percent on their premiums. But as *CNET* notes, "not just how many miles you drive but also how you're driving them could affect your insurance rates.[52]"

BEYOND FITNESS

While behavior change might not be your ultimate goal, Automatic Link leverages the OBD-II post in your car to report data back to your smartphone via Bluetooth. The device allows you to monitor your car's speed, fuel injection rate and more. There's also an accelerometer in the device, which detects sudden stops and starts and a speaker that beeps in those moments, as if the device were chiding you about your driving habits.

The app will alert you when you're going over 70 miles per hour or if you've accelerated too quickly after a stoplight. The idea is that you can learn more about your driving behavior, which is distilled into a single score each week.

As a bonus, the device can provide you additional information should your car display at Check Engine light and suggest ways to fix the issue. Also useful is a featured called Crash Alert which will telephone an emergency number and transmit your GPS location in the event of a crash[53].

This is exactly the type of data Nest thermostats brought to your home and has saved US customers on average about 10 to 12 percent on their heating bills and about 15 percent on their cooling bills.

Southern California Edison saw the opportunity to provide customers with additional insight into their utility bills following their

52. http://www.npr.org/sections/thetwo-way/2015/04/08/398306430/john-hancock-hopes-youll-trade-activity-data-for-insurance-discounts
53. http://techcrunch.com/2013/10/18/automatic-link-review-or-how-i-learned-to-stop-worrying-and-love-realtime-driving-feedback/

rollout of smart meters. In one test, the company identified 30,000 customers whose bills were on track to be significantly higher than expected. Ten days into the billing cycle, the system sent an email with the header, "Your bill is going to be higher than you expect, and we're concerned." More than 50 percent of customers opened the email. Compared to a control group, satisfaction rose by double digits, energy usage fell, and customer calls decreased[54].

The lesson here is that customers love to learn more about themselves and especially through data. This empowers them to change behaviors that may be costing them money or making them less healthy. Companies who offer this type of empowerment are in a great position to improve customer satisfaction and their bottom line.

IMPROVED CAREGIVING

Keeping track of one's own live is often complicated, much less the additional strain of caring for a sick or elderly loved one.

Express Scripts piloted a smart pill bottle, which proactively reminded patients to take their medication. The container, which featured a top called a GlowCap, connects to a wireless transmitter you place in your loved one's home.

When it is time for a dose of medicine, the GlowCap emits a pulsing orange light; after an hour, the gadget starts beeping every five minutes, in beeps that become more complicated and insistent. After that, the device can set off an automated telephone or text message reminder to patients who fail to take their pills. It also can generate email or letters reporting to a family member or doctor how often the medication is taken[55].

54. http://www.baselinemag.com/crm/how-technology-contributes-to-customer-experience.html
55. http://www.wsj.com/articles/SB10001424052748703431604575095771390040944

Tricella has unveiled a similar solution, which is a smart pill box. Families can help their loved ones fill the box and receive a notification if they do not take their pills. This allows family members to monitor their loved ones from a distance and send them a reminder when needed[56].

Lively has rolled out an entire line of sensors to help monitor aging or ill loved ones. In addition to a watch-like device to monitor motion, they produce a kit with pillbox and door sensors. Communicating with the Lively hub, family members can be alerted in the event of irregularities. Much like the Life Alert pendants, the Lively watch features a button which allows the wearer to summon emergency aid[57].

Taken together these fitness, automotive, energy and healthcare innovations have should serve as a lesson to your business. From our perspective, they offer three benefits to consumers:

1. Automation - users don't have to actively track data
2. Insights - users can learn more about themselves or others
3. Empowerment - with data comes the ability to act, whether that's changing behavior or taking another action

MEDICAL ACCESS

While most developed countries enjoy the benefits of modern healthcare, undeveloped countries often fall woefully behind. Zipline stands out as a beacon of innovation, merging the worlds of AI, data mining, and drone technology to revolutionize medical supply delivery. Imagine a world where life-saving medical supplies can be delivered within minutes, cutting through the

56. http://www.tricella.com/
57. http://www.usatoday.com/story/tech/columnist/baig/2015/02/05/review-lively-safety-watch-isnt-perfect-but-it-could-save-a-life/22925659/

constraints of terrain and distance. That's the world Zipline is creating. By deploying a fleet of drones, they ensure that essential medical products, like blood, vaccines, and medications, reach remote or hard-to-reach areas, especially in parts of Africa where road infrastructure can be a major barrier to timely healthcare delivery.

At the heart of Zipline's operation lies a sophisticated AI and data mining system. This isn't just technology for technology's sake; it's a carefully orchestrated dance of algorithms and analytics. The AI system pores over vast amounts of data, including weather patterns, geographic information, and healthcare facility needs. It's like having a super-smart logistics coordinator that can predict and plan the most efficient routes, ensuring that these vital deliveries are not just fast but also reliable. The AI doesn't stop learning - it continuously refines its analysis, ensuring that each flight is safer and more efficient than the last.

Zipline's use of AI and data mining exemplifies how technology can be harnessed for profound social impact. It's not just about the drones whizzing through the sky; it's about the lives saved and the healthcare systems strengthened. In areas where Zipline operates, the availability of medical supplies is no longer dictated by the harshness of the terrain or the unpredictability of the roads. It's a testament to the power of AI when it's driven by a mission to do good, blending cutting-edge technology with a deep commitment to improving human lives. In the narrative of technological advancement, Zipline's story is a heartening reminder of the potential for tech to be a force for good.

FOUR ADDITIONAL EXAMPLES

Here are four more examples that showcase how various companies and technologies leverage personal data and behavior change to enhance customer experience and engagement:

1. **Peloton's Personalized Fitness Classes:** Peloton utilizes user workout data to recommend classes that align with their fitness level and goals. This approach encourages continuous engagement and improvement by providing tailored workout recommendations.
2. **MyFitnessPal's Nutritional Insights:** MyFitnessPal tracks users' dietary intake and exercise to provide personalized nutritional advice. By offering insights based on personal health data, it allows users to make informed decisions about their diet and fitness routines.
3. **Grammarly's Writing Assistance:** Grammarly analyzes users' writing to offer personalized corrections and suggestions. By learning from users' common mistakes and writing style, it provides targeted advice to improve their writing skills.
4. **Fitbit's Health and Fitness Tracking:** Fitbit uses data on physical activity, sleep patterns, and other health metrics to offer personalized insights. This encourages users to lead healthier lives by providing tailored recommendations and tracking progress towards their health goals.

All these examples showcase how leveraging personal data can significantly enhance user engagement and provide tailored experiences in various domains, from health and fitness to writing and e-commerce.

As we mentioned at the beginning of this chapter, we believe companies who use personal data to empower their customers have an opportunity to create transformative change for those customers. With empowerment and then success, gratitude and loyalty are often the next set of emotions. There are no greater emotions you can have associated with your brand.

PART III:

THE SECOND R

CHAPTER 7

RESPONSIVENESS

"It's not about the big bang; it's about constant responsiveness."

– Charlene Li and Josh Bernoff

In our on-demand world, customers expect companies to respond to their needs. According to research by IBM, four out of five customers demand improved response time from businesses they frequent. Consumers no longer judge you in comparison to the competition, they are holding you up to the bar you set in their latest experience.

MAYDAY, MAYDAY

Among companies today, Amazon is arguably the leader in continuously raising the bar. An example of this was its Mayday button for the Kindle Fire HDX tablet. The single-click support solution allows tablet owners to access an Amazon customer service representative via webcam. Within 15 seconds, an Amazon representative shows up on-screen to solve problems in a responsive fashion with a simple touch of a button.

Amazon founder Jeff Bezos says it's, *"actually very similar to having someone standing next to you."* In addition to the live video, Mayday call reps can draw lines and shapes on the user's screen to provide visual instruction, telling the user which buttons to press or which menu to open. They can even control the tablet directly if the owner has enabled the feature.

According to *TechCrunch*, "[Amazon] built a full infrastructure to support it at their HQ in Seattle and on board the hardware. By compressing the video signals, they can send more data to the devices from tech support and allow tech support to see the data remotely."

Providing tech support isn't an easy proposition for both the company and the customer. The Mayday system helps both parties. Customers get better customer service while Amazon reduces the costly number of unnecessary phone calls. It's convenient too. Mayday is online 24 hours a day, every day of the year, with live representative available at any time.

The feature has been a huge hit with consumers. Among tablet owners, 75 percent of customer service interactions now come via the button. Amazon has now extended the service to its Fire phone[58]

Why did this work so well? It's because Amazon is responding to its customers at the exact time of need. It's leveraging technology to cure little problems before they become big thorny ones. The average time it takes for the representative to pop up on the screen is under 10 seconds[59].

A couple of FAQs on Amazon's website tell the story of how Amazon is focusing on leveraging technology as a Blue Goldfish to set themselves apart from the competition:

> **Is this your attempt to compensate for not having a genius bar?** No. This is a completely new approach to tech support: you get your own personal tech advisor from the comfort of your home or office.
>
> **This seems like a weird thing to invest so much time and money into?** We strive to be Earth's most customer-centric company, and this is just one more way we are working to achieve that goal.

Earth's most customer-centric company has set the bar on responsiveness. As with all expectations, you can't meet them. You can either exceed them or fall short. The choice is yours.

---

58. http://techcrunch.com/2013/09/24/amazon-introduces-mayday-a-unique-and-a-mazingly-useful-live-tech-support-system-for-kindle/
59. http://www.techtimes.com/articles/8557/20140617/amazon-mayday-response-ti-me-is-9-75-seconds-bad-news-for-kindle-fire-hdx-rivals.htm

I JUST WANT THE RIGHT CLOTHES

In the bustling world of retail, American Eagle Outfitters is leading a remarkable revolution, reshaping the way we shop with a blend of innovation and technology. At the heart of this transformation is the company's state-of-the-art mobile app and AI fitting room technology, both designed to cater to the modern consumer's desire for speed, convenience, and personalization. Imagine stepping into a store where your shopping experience is uniquely tailored to you – that's the promise of American Eagle's forward-thinking approach.

Here's how it works: The mobile app transforms your smartphone into a personal tailor. With a simple scan or entry of your measurements, the app precisely calculates your body dimensions. This isn't just about getting the size right; it's about creating a shopping experience that feels custom-made. The app curates clothing suggestions and recommendations that are perfect for your body type, turning the often-daunting task of finding the right fit into a delightful experience. The magic doesn't stop there. American Eagle's interactive fitting rooms are like something out of a sci-fi novel. Gone are the days of peeking out from behind curtains, searching for a different size. Now, with the touch of a screen installed in each dressing room, you can scan items, access product information, and even summon different sizes and styles directly to your fitting room. It's shopping convenience at its finest.

But the benefits of this technology extend beyond the customer. Imagine being a store associate in this high-tech environment. When a customer requests a different size from the comfort of their fitting room, associates receive instant notifications on their mobile devices. They can respond swiftly, streamlining the in-store logistics and ensuring that customers' needs are met promptly and efficiently. This synergy between technology and human touch not only enhances the shopping experience but also revolutionizes the way store associates operate, making their work more efficient and enjoyable.

American Eagle's visionary approach is reshaping the retail landscape. By embracing digital innovation and integrating it seamlessly with the traditional shopping experience, they're not just keeping up with the times; they're setting a new standard. In today's fast-paced world, where customer expectations are ever-evolving, American Eagle's commitment to a seamless, unified commerce experience is more than just a business strategy; it's a testament to their dedication to meeting and exceeding the needs of their customers. This isn't just about shopping; it's about experiencing the future of retail, today.

As we draw the curtain on our exploration of customer service's responsiveness, it's undeniable that the burgeoning role of Artificial Intelligence is reshaping the landscape in profound ways. This chapter isn't just about technology; it's a testament to the extraordinary partnership between human empathy and AI's analytical brilliance, a partnership that's reinventing how we connect with and serve every customer in a timely manner. The challenge for us isn't merely adapting to AI, but evolving alongside it, crafting responsive experiences that don't just satisfy, but truly resonate with those we aim to serve.

CHAPTER 8

CUSTOMER SERVICE 3.0

"We see our customers as invited guests to a party, and we are the hosts. It's our job every day to make every important aspect of the customer experience a little bit better."

– Jeff Bezos

Many companies talk about superior service and the importance of going the extra mile for customers. How many though actually walk the talk? To steal an immortal line from former Texas Governor Ann Richards when describing George W. Bush, customer service is typically, "all hat, no cattle." Or more aptly, "all PowerPoint, no action."

There are some companies leveraging AI, data, and analytics to deliver a next level of customer service. A level we call 3.0.

Eileen Scully wrote about a heartwarming story from her friend Michael John Mathis. Michael had posted the following message on *Facebook* recounting an episode of lost glasses on a train:

> I just experienced possibly the very best customer service of all time. A few weeks ago, I accidentally left my beloved Warby Parker reading glasses on the Acela. Annoyed, I bought myself another, identical, pair the following day. Today I received an unexpected package containing not one, but two pair of those same reading glasses, a copy of "*On the Road*" by Jack Kerouac and the following note:
>
> > *'Hi Michael, This might be odd... but you sat across from me on the train ride from NYC to Boston a few weeks ago and left your glasses on the train! As luck would have it, I happen to be the GC of Warby Parker, and there is nothing I like more than a good mystery... I hope these find you in good health! (also, we noticed your lenses were scratched so we made you a fresh pair!) Sincerely, AK'*
>
> I am so impressed! What a remarkable person and company. They have a customer for life!

AK is short for Anjali Kumar. A few points to share about Anjali and Warby Parker. Warby is a B Corporation based in New York.

Like TOMS Shoes, they give away one pair of glasses for every pair purchased. The name Warby Parker is derived from two different characters created by Jack Kerouac: Warby Pepper and Zagg Parker.

So, how did Anjali find Michael John? Of the 45 comments on his Facebook post, the majority focused on how impressive the act was. The superlatives included: GREAT! WOW! INCREDIBLE! AMAZING! UNBELIEVABLE! and "And from a lawyer even! Go figure!" But there were a few of the comments that focused on how this happened.

Here's an exchange between Michael and one of his friends:

> **Friend:** THAT'S AMAZING! I've heard nothing but great things about them. My co-worker – Mr. Skeptical had 20 questions though – like how did he get your name and mailing information?
>
> **Michael John Mathis:** NR, I imagine it was not that tough – just very kind. Not that many pair of that style, in that exact prescription, sold to a guy in NYC, plus I ordered another pair the following day online. That plus a quick check of a photo on Facebook or LinkedIn to be sure I had been the guy sitting across from her on the train.
>
> **Friend:** SO AMAZING! My co-worker is young and fearful of those that do good – he still has 20 questions and I'm like are you kidding me can't you just accept that there are good people in the world.! NICELY DONE AK OF WARBY PARKER!!!!!!!

There's a reason we call this 3.0 and it's the same reason Michael's friends were impressed: Warby Parker chose to give more. More effort. More value. More responsiveness.

We don't just like this example because Warby Parker demonstrated heroic levels of responsiveness. We like it because it wouldn't be possible without technology. Without a comprehensive record of customers and orders in a searchable format Anjali, even with his desire to go above and beyond, would've been stymied by the time required to find Michael's order. In some companies accessing data in this way would've been next to impossible.

As you read this chapter on next-level customer service, realize that each of these interactions was enabled by having the right combination of people and technology in place. Without both achieving heroic levels of responsiveness will prove elusive.

SPEED MATTERS

When we use the word responsiveness, it's usually to describe someone who responds quickly. By this point, you know customers expect increasingly faster response times. In fact, Facebook now includes a special indication for brands who respond to inquiries quickly.

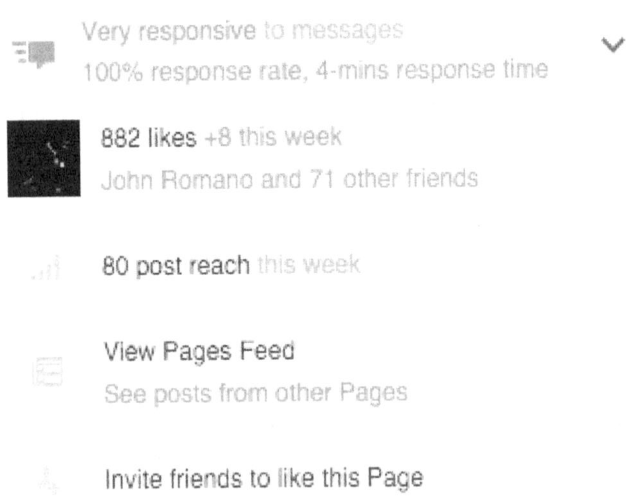

Brands can enable this badge by responding to 90 percent of messages in five minutes or less. For a brand with a small following, this isn't difficult. But how do you have a five-minute response time when you're receiving thousands of messages a day, much less come close to Amazon's 15-second Mayday Button? The answer lies in technology.

Speed Matters at Groupon

Just after its fifth birthday, Groupon's social fan base of 15 million was generating nearly 85,000 inbound messages a week. They were responding to less than 15 percent and were soon swamped. Groupon needed a way to manage consumer relationships, at scale.

The "One Playbook" was born. The initiative bridged across all facets of the organization, including a complete revamp of Groupon's social media strategy. The rollout of One Playbook created a global approach to content production, distribution, and measurement for Groupon's social team. This required building a social relationship infrastructure that leveraged technology to scale their social efforts, while building and nurturing customer relationships at a local (and personal) level.

Groupon's journey stands as a testament to adaptability and the relentless pace of market evolution. Back in its heyday, Groupon was a paragon of responsiveness and customer engagement. Its "One Playbook" initiative was a groundbreaking strategy, setting new standards in managing consumer relationships at scale. The program's success was evident in its staggering engagement growth and the rapid response to customer inquiries, making Groupon a model for effective social media strategy and customer relationship management.

However, the relentless advance of e-commerce has reshaped the playing field, presenting new challenges that have tested Groupon's once-dominant position. The surge in online shopping, propelled by behemoths like Amazon and eBay, has nudged Groupon from its comfortable perch. These e-commerce giants, with their vast inventories and aggressive pricing strategies, have encroached on Groupon's territory, offering consumers an alternative to the high-discount coupon deals that were once Groupon's forte.

This shift in consumer preferences and the intense competition has compelled Groupon to rethink and realign its business model. The company's pivot towards local mobile commerce and direct merchandise sales is a strategic response to the evolving market dynamics. Groupon's transformation reflects the broader industry trend where agility and innovation are essential for survival and growth.

While Groupon's responsiveness and customer engagement once set industry benchmarks, the company now finds itself in a transformed e-commerce landscape. This new environment, characterized by heightened competition and changing consumer behavior, has necessitated a strategic shift for Groupon. The company's ability to adapt and innovate will be crucial in maintaining relevance and competitiveness in this ever-evolving digital marketplace.

Agility is also key. But what happens when the core of your business model finds itself adrift in the relentless tides of change? Responsiveness alone can't steer the ship back to success, as we've seen with companies like Groupon, struggling to navigate the shift to e-commerce.

Consider the contrasting journey of Ebags.com, an online retailer that's become a beacon in the sea of bags, luggage, and travel accessories. Their commitment goes beyond just offering an extensive range of products or competitive pricing. At the heart of Ebags.com's success is a customer service that doesn't just respond; it anticipates and adapts.

Picture this: tens of thousands of orders processed, not a single complaint lodged. That's the hallmark of Ebags. It's no small feat, and it's not gone unnoticed. They've earned an "Elite" rating from Stella Service, a testament to their exceptional customer care. When you call their representatives, it's not just a conversation; it's a connection.

This narrative of Ebags isn't just about their responsive customer service; it's a testament to the power of responsiveness in an ever-changing consumer landscape. It's a story of how staying attuned to customer needs and trends can transform a business, propelling it to new heights even as the market shifts beneath its feet.

Ebags stands as a compelling example, demonstrating that true responsiveness in business isn't just about speed; it's about understanding, adapting, and thriving amidst the tides of change.

TOOLS TO BETTER KNOW CUSTOMERS

Speed, however, is only one element of responsiveness. If we look to Webster's, to be responsive is to "respond especially readily and sympathetically." So, we not only must respond with speed, but we must also do so understanding our customer.

Have you ever rushed into a meeting, not having a chance to do any research on who exactly it is you'll be talking to? In the words of Jason Cipriani of *CNET*, "I know I have, and it almost always leads to an awkward, forehead-slapping 'I should have known that' moment.'"

Enter Clara, an innovative AI scheduling assistant that not only helps manage your calendar but also provides key insights into the people you'll be meeting. Unlike traditional apps, Clara operates through email, seamlessly integrating into your existing workflow. Once you schedule a meeting, Clara springs into action, gathering essential information about the attendees and the companies they represent. Imagine being notified about a recent achievement or notable event related to the company of your meeting attendees

– Clara can bring these details directly to your inbox, ensuring you're always well-informed and ready to make a positive impression.

Similarly, CalendarHero steps up as a powerful tool for meeting preparation. With its advanced scheduling capabilities, this app streamlines the process of setting up meetings. But it's not just about finding a convenient time; CalendarHero also integrates with your technology stack, enhancing your ability to prepare for meetings effectively. Whether it's pulling up relevant documents, providing context on attendees, or even reminding you of key talking points, CalendarHero ensures that you are never caught off-guard.

CalendarHero leverages artificial intelligence (AI) to streamline the meeting scheduling process by analyzing participants' calendars to find optimal meeting times, considering preferences and past scheduling patterns. This AI-driven approach enables the system to offer smart suggestions for meeting slots, reducing the back-and-forth communication typically required to coordinate schedules. Furthermore, CalendarHero's AI capabilities extend to automating various administrative tasks, such as setting reminders, sending follow-ups, and integrating with other digital tools to enhance productivity. By employing machine learning algorithms, CalendarHero continuously improves its scheduling recommendations and efficiency, adapting to users' changing needs and preferences over time. This intelligent utilization of AI makes CalendarHero a sophisticated solution for managing meetings and workspaces in today's fast-paced business environment.

The more informed you are about the people and companies you interact with, the better you can connect and communicate. Both Clara and CalendarHero represent the next step in intelligent meeting preparation, equipping you with the tools to respond quickly and empathetically to your customers and colleagues. With these apps, those forehead-slapping moments are a thing of the past.

And soon with the advent of advanced facial recognition the concept of wearable glasses that can easily identify the person in front of you will be reality. In her eye-opening book on facial recognition – "Your Face Belongs to Us: A Secretive Startup's Quest to End Privacy as We Know It" author Kashmir Hill shares her experience with Clearview Ai's technology. It is not available to the public, but as goes with science fiction, soon what we see in a movie becomes reality.

Understanding Customers to Personalize Verbal + Written Communication

Crystal Knows is an app that analyzes public data to tell you exactly how to communicate with people. It delivers information that's eerily accurate. The app touts itself as the "best improvement to email since spell-check and it might not be entirely wrong. Crystal scrapes publicly available information about you and others to create a profile on how to best communicate with you, complete with details about whether or not you like emoji or sarcasm.

According to Owen Williams at The Next Web, "Take me, for example. Crystal says that I am "friendly, casual, and extremely perceptive, "connecting the dots" more quickly than others but occasionally rambling in conversation." I ramble, it's true. How the hell does it know this, though? It also suggests the best ways to interact with me — meetings over food, again, accurate — and some topics to avoid."

How does it work? Crystal grabs data about you in a few interesting ways. First, it asks people who visit a person's profile to fill in information about them and scrapes "thousands of sources" online for as much data as it can find. Second, the service basically stalks you online to find out how you talk to others and the results they manifest. You can plug Crystal into Gmail for quick tips about whoever you're emailing.

Tools like Clara give people information about each other before they even meet, and Crystal Knows is a service that analyzes public data to tell users how to best communicate with specific individuals via email and verbal correspondence[60].

CONTEXT IS KING WITH RESPONSIVENESS

Ex-Googler Falon Fatemi wrote for *TechCrunch*, "The future of the Internet is all about context, and that future isn't far away. There's already been an uptick in context-rich systems hitting the market that allow data to be tailored to specific customer situations… Everything in tech is moving toward predictive pre-search, understanding user intent, and surfacing relevant info based on this data.[61]"

This trend holds good and bad news for companies today. The bad news is that these context-aware technologies will drive customer expectations higher. The good news is that the same technologies will be available to you as well.

Tools like Crystal Knows and Clara allow companies to know more about their individual customers. We anticipate more of these context-aware tools coming to market in the coming years.

Clear evidence of this trend is IBM's acquisition of business-to-business assets from The Weather Company, the parent company of The Weather Channel. From the press release announcing the acquisition:

> The Weather Company's extremely high-volume data platform, coupled with IBM's global cloud and the advanced cognitive computing capabilities of Watson, will be unsurpassed in the Internet of Things,

60. http://thenextweb.com/apps/2015/04/15/creep-on-your-friends/
61. http://techcrunch.com/2015/08/18/the-future-of-the-web-is-all-about-context/

> providing our clients significant competitive advantage as they link their business and sensor data with weather and other pertinent information in real time.

The implications for business are significant. While it may seem obvious to say, weather affects business. Energy companies care about avoiding outages. Insurance companies want to anticipate and reduce claims. Governments must handle emergency response. Retailers want to understand seasonal demand.

Context may be king, but costs and profitability rule! IBM decided to sell the assets of The Weather Company, including The Weather Channel's digital consumer-facing offerings like Weather.com and mobile apps, to Francisco Partners is part of a strategic move to streamline its operations. The transaction includes The Weather Company's various digital properties and enterprise offerings across industries like broadcast, media, aviation, and advertising technology.

The Weather Company, under IBM's stewardship, expanded to serve over 415 million people monthly through its consumer-facing properties and over 2,000 businesses with its enterprise products. However, IBM has been refocusing its business strategy towards its core offerings in hybrid cloud and AI technologies.

Francisco Partners, the acquiring firm, sees a promising future for The Weather Company. The firm intends to expand investment and resources in The Weather Company, enhancing its capabilities beyond weather forecasting. The aim is to develop new tools and experiences, particularly focusing on health and well-being, and to provide more actionable insights for businesses and consumers.

What is one company's data set and related cost is now another company's treasure. Data is key, but with the proliferation of AI connected to vast amounts of data, the question moving forward becomes – is data the most important part of meeting customer needs or is it possible that AI can extract existing data and move the needle forward to anticipate customer needs and solve them?

This is one additional data point to understanding a customer's context and tailoring your response. Doing this faster and even before the customer realizes they have a need is the new Holy Grail of customer service—anticipating needs.

ANTICIPATING AND ADDRESSING NEEDS

A Harris Interactive survey of more than 2,000 consumers revealed three takeaways about the future of customer service:

- 87 percent of U.S. adults are receptive to being proactively contacted by an organization or company when it comes to service and support.
- 73 percent of those surveyed who had a pleasant surprise or positive experience with proactive communication from a brand, report they also had a positive change in their perception of that organization.
- 62 percent said they acted because of that positive experience.

As brands adjust to the emergence and convergence of both new and more customer service channels, reactive customer service remains the day-to-day norm for most organizations. Forrester Research's Forrsights Survey shows that currently, only 29 percent of enterprises are investing in proactive outbound communications, even though customer expectations for real-time updates and information are steadily increasing.

Companies need to shift from predominantly reactive to reactively proactive. The difference can greatly impact customer satisfaction and retention. Done correctly like in the case of Amazon Mayday, it becomes a differentiator for the brands that can (even at times) delight in this manner.

According to Parature, companies can towards more proactive customer service by leveraging alerts and updates. They recommend using channels such as mobile, social, IVR messaging, and knowledgebase/FAQ updates to provide reactively proactive or proactive information and updates about service and product statuses.

For example, utilities companies can use the above-mentioned channels to proactively alert and update customers on outages and estimated service restoration times. Retailers can alert customers as to shipping specials, price reductions, product availability or recalls, and airlines can proactively update passengers regarding delays, cancellations, or weather announcements.

An example of this type of proactive service and support is Delta Air Line's mobile app which proactively delivers flight notifications and allows customers to rebook without having to call or wait in line at the airport. All of these proactive communications can reduce high call or email volumes as well as social media complaints.

Taking proactive service a step further, Bell Canada mines its customer interaction data to understand the relationships among various customer issues. Using what it learned about "event clusters," Bell began training its reps not only to resolve the customer's primary issue but be responsive to anticipate and address common downstream issues. For example, a high percentage of customers who ordered a particular feature called back for instructions on using it. The company's service reps are now proactive by giving a quick tutorial to customers about key aspects of the feature before hanging up. This allowed Bell to reduce its "calls per event" by 16 percent and its customer churn by 6 percent. For complex downstream issues that would take excessive time to address in the initial call, the company sends follow-up emails—for example, explaining how to interpret the first billing statement. Bell Canada is currently weaving this issue-prediction approach into the call-routing experience for the customer[62].

Money or Time-Saving Suggestions

While predictive analytics have been a topic of conversation for many years now in customer service, the technology to make full use of them is now catching up.

Google has been a front-runner, constantly pushing the boundaries of what's possible. A shining example of this relentless innovation is the transition from Google Now to Google Assistant. Google Now, introduced in 2012, was a groundbreaking step in AI, providing users with a virtual assistant that offered information proactively, based on their habits and preferences. It was like having a personal assistant who knew you so well that they could predict your next move.

However, technology, much like time, never stands still. Google Now, with its card-based interface and reliance on user input, was just the beginning. Users craved interaction, a conversational partner rather than a silent assistant. Enter Google Assistant in 2016, a significant upgrade in Google's AI journey. This wasn't just an update; it was a complete overhaul. Google Assistant was designed to be more than just responsive; it was built to engage, to converse, and to understand the nuances of human communication. It was like moving from a one-way street to a bustling highway of two-way interaction.

62. https://hbr.org/2010/07/stop-trying-to-delight-your-customers

The leap from Google Now to Google Assistant is not just about smarter technology; it's about creating a more human experience. Google Assistant's ability to engage in two-way conversations, understand context, and even showcase a sense of humor brought AI closer to the realm of human-like interaction. This transition symbolizes a shift from AI as a tool to AI as a companion. Whether it's managing tasks, controlling smart home devices, or simply sharing a joke, Google Assistant makes AI interaction feel more natural and intuitive.

Reflecting on this evolution, it's clear that the journey from Google Now to Google Assistant is not just a technological leap; it's a reflection of our own human desire for connection and interaction. As we look to the future, one can only imagine what the next chapter in AI will bring. Will it be more personalization, deeper understanding, or perhaps an even more seamless integration into our daily lives? One thing is certain: the journey from Google Now to Google Assistant is just the beginning of a much larger story in the world of artificial intelligence.

Gemini represents Google's exploration into advanced AI language models, like OpenAI's ChatGPT. It's focused more on generating conversational responses and providing information or creative outputs based on a wide range of internet knowledge. While Gemini (formerly Bard) uses sophisticated AI technology, its main function is different from that of a personal assistant.

The development of Gemini indicates Google's continued investment in AI and its commitment to exploring new applications of language models. It's a sign of the expanding capabilities of AI in understanding and generating human-like text, but it doesn't replace or directly iterate upon Google Assistant. Instead, Gemini and Google Assistant can be seen as complementary technologies, each serving distinct purposes within Google's broader AI strategy.

Matchmaking Improves Customer Service

"Your call may be recorded for training and quality purposes." We've all heard this before. Companies use call center recordings for training and compliance purposes. But imagine a solution that uses technology, data, and analytics in real-time to automatically match customers to like-minded customer service representatives. Software that uses algorithms to analyze customer speech for tone, tempo, and syntax to identify the customer's personality type and behavioral preferences.

Mattersight has developed exactly this. A solution that uses millions of pre-built algorithms to automatically decode every second of captured human interaction. Called Predictive Behavioral Routing, Mattersight can easily and instantly route each caller to agent who has demonstrated a high level of rapport with customers of that type. When that match is made, agents can more easily and more genuinely help customers. No more square pegs in round holes. The whole process becomes more efficient, and without a doubt, this makes for a better experience for customers.

According to Jason Wesbecher, CMO of Mattersight, "When you think of big data, there's a ton of data in that spoken conversation. No one is taking that data and trying to operationalize it in a new way."

The benefits according to Forrester are four-fold:

1. Incremental revenue
2. Reduced customer service costs and effort
3. Improved customer satisfaction
4. Reduced hiring costs due to a decrease in turnover

Not only does this tool improve the customer experience, but it enhances the employee experience, as well. The software is currently being used by major retailers and insurers, including CVS and Esurance. When reps are paired with these matched customers, they're easily able to communicate with, there's an increase in positive interactions, which translates to a greater sense of pride because they can perform their jobs better.

The bottom line is that people buy from people; and people especially like to buy from—or interact with—people they like and connect with[63].

In recent years, Mattersight has released updated versions of its applications. The latest, Version 3 of its Predictive Behavioral Routing algorithm, was designed to improve customer experience by enabling organizations to have shorter and more efficient conversations with customers. This version purportedly advances client key performance by 20 to 25 percent. The algorithm connects customers and agents based on their interactions, routing calls to the most suitable agents, which significantly improves customer satisfaction and service efficiency.

Additionally, Mattersight introduced Workstyle, a lightweight SaaS app that trains call center employees to enhance customer experiences by improving personality connections. Workstyle complements the Predictive Behavioral Routing by leveraging the power of personality in service, sales, retention, and collections interactions. It provides users with personalized learning pathways, helping them to identify, understand, and better connect with different personalities they encounter daily.

Mattersight's suite of personality-based software applications has substantially influenced customer interactions, resulting in shorter, more satisfying experiences. The company's mission is to facilitate more effective and effortless conversations between brands and

63. http://www.cx-journey.com/2015/11/customer-service-matchmaking.html

their customers, utilizing its patented stack of SaaS applications to predict customer behavior based on the language used during service and sales interactions.

FOUR ADDITIONAL EXAMPLES

Here are four additional examples of innovative Customer Service 3.0, showcasing how companies across different industries leverage technology to enhance customer service:

1. **Zappos' 365-Day Return Policy and Customer Support:** Zappos sets a high standard for customer service with their 365-day return policy. They also have a 24/7 supportive customer hotline. This commitment to customer satisfaction, alongside their empowerment of customer service representatives to make independent decisions to solve customer issues, has made them a benchmark in the industry for exceptional customer care.
2. **Ritz-Carlton's Empowerment of Employees:** Ritz-Carlton empowers its employees with the ability to spend up to $2,000 to solve any guest issue without needing managerial approval. This empowerment is part of their legendary commitment to guest satisfaction, allowing for immediate problem resolution.
3. **Bank of America's Erica, the Virtual Financial Assistant**: Bank of America introduced Erica, a virtual financial assistant that uses predictive analytics and cognitive messaging to provide financial guidance to more than 45 million customers. Erica can help with everything from transaction queries to bill payments.
4. **Uber's Real-Time Feedback Loop:** Uber's service includes a real-time feedback loop where riders rate their drivers immediately after their ride. This system ensures high standards of service and safety, as drivers with consistently low

ratings are removed from the platform. It exemplifies how technology can be used to maintain service quality and address problems promptly.

All of these examples illustrate how companies are leveraging technology, personalized services, and employee empowerment to push the boundaries of traditional customer service, ensuring memorable and satisfactory customer experiences.

ONE STEP AHEAD OF THE CUSTOMER

We said previously the only way to compete is to get ahead of the curve and offer service at a level greater than a customer's current expectations. There's no doubt that customers don't just expect you to address issues as they arise, rather they'll soon expect you to help them with issues they don't even know they have.

Whether you're re-booking a flight before the customer knows it's cancelled or identifying which students are at risk of not graduating, as Arizona State University does,[64] data, real-time analysis and immediate response of Customer Service 3.0 is going to be the new normal.

64. http://www.nytimes.com/2014/12/03/opinion/blowing-off-class-we-know.html

CHAPTER 9

WAITING

"And sure enough even waiting will end...
if you can just wait long enough."

– William Faulkner

Someone once said we spend 10 percent of our lives waiting. We're certain it's true because we once read it online. In all seriousness, waiting is a fact of life. A rather irritating, annoying fact of being a customer.

One of the most impactful ways companies are using technology is to help their customers wait less, or when waiting can't be avoided, help the time pass more quickly.

Imagine this situation. You've just arrived on a late flight, found your way to your hotel and walk through the lobby queue to a lengthy queue of guests waiting to check in. Not only does a five-to-ten-minute wait seem long, but you're also tired and growing more irritable by the second. We've all been there and waiting, especially in this situation, is exactly the last thing you want to do.

Both Marriott and Hilton have rolled out solutions to prevent this from happening. Replacing plastic key cards with Bluetooth technology, guests can now unlock their room with their smartphone.

The apps offer an option to enable mobile check-in on its home screen. For those who might miss it, a push notification reminds travelers to sign up before their trip. Setup is simple and requires users to submit their credit card information, so that front desk transactions can be bypassed entirely. It takes roughly a minute from start to finish.

As guests make their way to their destination, relevant notifications will tell them whether they've been upgraded or if their room is ready, at which point they can request a mobile key with the tap of a button. For privacy reasons, the room number will appear only when you ask for it; a diagram on the same screen offers easy instructions for how to make your phone communicate with the door's sensor. According to Nikki Ekstein, "When I tried it out at the W Downtown, in New York's Financial District, I was able to through the lobby and head straight to my room—no human contact necessary."

Leveraging self-service technologies to reduce waiting, doesn't just save customers time, many genuinely prefer to help themselves. It's simply empowering.

Hertz knows this fact well. Hertz on Demand enables renters to unlock and lock their vehicles using RFID. Hertz on Demand members receive a key fob, and it comes with an embedded RFID reader. After receiving a confirmation email and being assigned a vehicle, members can swipe the key fob across an RFID reader placed on the lower left-hand-side of the car's windshield. If they have found the assigned vehicle, the authorized user ID is transmitted to the car's on-board computer and the rental gets activated and the doors unlocked. The engine also is engaged, and members can use the keys inside the vehicle to get their road trip under way. When the rental is over, members press a button inside the vehicle to indicate they are finished. Quick and painless[65].

COFFEE ON THE GO

Of course, not everything can be completely self-service, however most any company can benefit from considering which parts of their experience can be augmented with technology to make the process go more smoothly for customers.

Whether you prefer Starbucks or Dunkin' for your coffee, both now serve as an example for how you can avoid waiting. Mike Wittenstein had this to say of the experience:

> Starbucks wasn't the first place to let customers buy something online and pick it up in-store. However,

65. http://www.tnooz.com/article/futuristic-rfid-is-here-and-now-for-hertz-on-demand-and-zipcar/

> they've done a good job of integrating the new service into their operations. The details have been engineered into an amazingly simple experience that makes it easy on customers—and easy on the baristas.

SELF-SERVICE IS THE NEW FULL SERVICE

Countless other companies are embracing the self-service trend. Julia Stewart, CEO of Applebee's pulled the trigger to buy 100,000 tablets for restaurant tables. "Customers have been telling us for some time—even myself ... I don't like to wait for the check," said Stewart. "That was the first sort of pain point we heard of, and we had this unique opportunity with technology to make a real difference."

The result from Applebee's shows 70 percent of tables using tablets, which also increased appetizer orders by 20 percent and dessert orders by 30 percent. Sounds like a win-win as customers are leaving faster, more satisfied and with a higher check total.

Looking to recent survey numbers, these results should come as no surprise. "Almost a quarter of all millennials use self-service kiosks to avoid any sort of interaction with cashiers," according to Retale president Pat Dermody, whose company recently commissioned a survey on self-checkout.

Millennials like self-checkout – but they'd like it even more if they could use mobile payments there. While 20 percent of the survey's 1,000-plus U.S. adult respondents said they'd like to have a mobile payment option at self-service checkout kiosks, the number was 26 percent among those age 18 to 34, versus 16 percent for those over 34.

Millennials were also more likely to have used self-checkout (91 percent, versus 81 percent for age 35 and up) and more likely to say they liked self-checkout because they "don't like interacting with cashiers" (20 percent, against 12 percent for all those surveyed). And fewer of those younger shoppers said they often need help from a store associate when using self-checkout (37 percent, versus 47 percent of 35-and-olders). Still, 53 percent of all those surveyed want at least one store employee around to make sure self-checkout goes smoothly.

Other common reasons for choosing self-checkout included "I have a limited number of items" (72 percent), "there was no line" (55 percent), and "I prefer to keep my transactions and financial information private" (13 percent).

That millennial preference for self-checkout—and a higher than usual aversion to dealing with store associates—lines up with a recent Forrester Research study, which concluded that retailers who cater to millennials are wasting their time because millennials don't have much money to spend. That report recommended that retailers forget about chasing younger shoppers with beacons and push notifications to smartphones, and instead invest in equipment that helps stores stay open earlier and later without extra costs—such as self-checkout kiosks[66].

Yet, while the idea of self-service is attractive to a certain demographic the practical aspects don't always pan out as planned. For example, Walmart announced in 2023 that it will be removing self-checkout stations from thousands of stores over the next two years, replacing them with more traditional staffed registers. The decision aims to improve customer service, reduce theft, and address complaints about the self-checkout experience. This removal is planned for approximately 2,500 stores, with 1,000 locations seeing registers replaced in 2023 alone.

66. http://www.pymnts.com/news/2015/millennials-want-self-checkout-with-mobile-payments

BEYOND SELF-SERVICE

With so many different locations, Hertz has a wealth of data it can analyze to improve the customer experience. From emails, text messages, and web surveys, the car rental company can gather more information on their customers, understanding what makes them tick. Through this greater understanding, Hertz has worked to enhance the company's performance by improving the relationship between workers and the customer. One example comes from analyzing the data to adjust staffing levels at an office in Philadelphia. By making sure a manager was on hand during peak activity hours, customer questions and complaints could be resolved more quickly and effectively, essentially improving each customer's experience[67].

Many retailers have learned this lesson as well because a lengthy line also gives customers a chance to think more about their purchases and potentially change their mind on some items or on the entire order.

> Retailers like Apple and Home Depot recognized the need to eradicate this wait several years ago, providing employees with mobile POS devices that enable them check out customers from anywhere within the store if the customer is paying with a debit or credit card. Employees at Apple's retail stores have been armed with iOS devices for several years, enabling consumers to easily make purchases without waiting in line. In early 2011, Home Depot introduced their "First Phone" to allow customers to check out from anywhere within the store, also without having to wait in line.
>
> Nordstrom, the Seattle-based fashion and beauty retailer offering apparel, shoes, makeup, and other beauty products, is rolling out mobile point-of-sale

67. http://dataconomy.com/5-use-cases-of-big-data-improving-customer-satisfaction/

(POS) devices throughout their full-line stores, as well as in some of their Nordstrom Rack stores. These mobile POS devices, which is a modified iPod Touch with a merchandise scanner and credit card slider, allow employees to check out customers from anywhere in the store. The app on the device also provides Nordstrom's sales staff access to the company's entire inventory, which is useful when helping customers check if an alternative size or color is available elsewhere, without needing a register to look up that information.

Nordstrom has deployed over 6,000 of these devices throughout their 117 full-line stores, and at some Nordstrom locations, there are more mobile POS devices than regular registers. Colin Johnson, a spokesperson for Nordstrom, says that these devices are part of a larger plan for Nordstrom to help "provide a more technology enabled store experience." He notes that in 2005, Nordstrom began offering the option to ship merchandise directly to customers, and in 2009 the company integrated inventory with its online store. In 2010, Nordstrom then introduced Wi-Fi into stores to "make it easier for customers to stay connected in the stores by using their mobile devices to shop and to compare and learn more about merchandise."

Johnson explains that the goal of using the mobile POS for Nordstrom is really to "take care of customer anywhere in the store. We don't have to take you to the cash register, and instead, can do that right there with you on dressing room or when you're trying on shoes – and then you're on your way." He adds, "that kind of ability to increase speed and convenience is increasingly important."

Taming the queue isn't just of interest to retailers, it's important for service-based businesses as well. Impact Urgent Care realizes this and if you've ever sat in a doctor's waiting room for more than twenty minutes, you do as well. Enter the Virtual Waiting Room.

> When Linda Ratner and her husband opened Impact Urgent Care, the two emergency room medical professionals soon realized they couldn't keep working 12- to 14-hour days and provide patients with quality care. Because the clinic helps patients with an array of complaints -- from flu shots to many conditions that previously would require an ER -- consultations could take anywhere from a few to many minutes, making it difficult to correctly manage the growing influx of patients, both walk-ins and those who made appointments.
>
> To try to satisfy patients and manage the increasingly unbearable workload, staff first tried a manual system: As it got closer to patients' appointments, the office would phone them so the in-office wait would be shorter, said Ratner, a registered nurse and executive director of Impact Urgent Care. Still, the clinic had to fit in walk-in patients, and the approach was far from perfect.
>
> "We did that with people who walked in off the street as well. That worked well for about a year, but it had a lot of human error. If we got busy, we might forget to call a patient," she said. With plans to open five more clinics in San Antonio, Ratner knew the clinic needed an automated system, so she began her quest again, researching how heavily trafficked organizations such as the Department of Motor Vehicles kept waiting room volumes low. That search led to QLess, software

> that allows patients to "get in line" via cellphone, online, or via the phone (where a clinic employee inputs their names through the administrative version of the software). The clinic educates patients about the system and uses this remote waiting room capability as a differentiator.
>
> "When we open the new clinics, patients will see the queues on the Internet and will be able to see which one they'll want to go to, based on the queue ahead of them. It's been good for our staff, our patients love it, and we love it," said Ratner. "For us, it lets us know what the day's going to look like. For the patient, it gives them the ability to get their life back."
>
> After implementing QLess, patient satisfaction scores improved 20 percent, she said—at a time when volume grew to an average of 50 patients per day this year, compared with 30 per day[68].

Solutions to problems don't always happen in expected liner ways. Take for example the problem with baggage wait time at George Bush Intercontinental Airport in Houston. Faced with a high number of passenger complaints about long waits at baggage claim, the airport executives initially attempted to solve the problem by increasing the number of baggage handlers. This strategy reduced the average wait time to just 8 minutes, significantly below the industry standard. However, this did not lead to a decrease in complaints.

In response, the executives conducted an on-site analysis and discovered that passengers only took about a minute to walk from their arrival gates to baggage claim, followed by a 7-minute wait

68. http://www.informationweek.com/healthcare/patient-tools/eliminate-the-waiting-room/d/d-id/1316203

for their bags. To address this, they employed an unconventional strategy: they moved the arrival gates farther from the main terminal and routed bags to the outermost carousel. This change meant that passengers had to walk six times longer to reach their luggage. As a result of these adjustments, complaints about baggage claim wait times dropped to near zero.

This approach exemplifies a creative problem-solving method where the focus shifted from speeding up the process to managing the passengers' perception of wait times. It's a strategy that highlights the importance of considering the entire customer experience, not just isolated aspects of it.

FOUR ADDITIONAL EXAMPLES

Here are four more examples of how companies manage customer wait times innovatively:

1. **Disney's Genie+ Service:** Building on its legacy of innovation, Disney introduced the Genie+ service within its app, offering a more flexible alternative to the traditional FastPass system. Genie+ allows guests to select arrival windows for various attractions and experiences throughout the day, effectively minimizing wait times and enhancing the guest experience.
2. **Chick-fil-A's Mobile Ordering and Table Service:** Chick-fil-A uses its mobile app to allow customers to order ahead and avoid waiting in line. In some locations, after placing an order through the app, customers can choose to dine in, and their meal is served to them at their table.
3. **Southwest Airlines' Boarding Process:** Southwest employs a unique open seating policy, where passengers are assigned a boarding group and position based on their check-in time. This streamlined process minimizes time

spent waiting in line and it improves the overall efficiency of the boarding experience.

4. **Cedar Point's Access Pass System for High-Demand Rides:** To manage the wait times for its most popular attractions, Cedar Point amusement park introduced an Access Pass system. Guests obtain a pass for free on a first-come, first-served basis. The pass assigns them a specific time window to return and enjoy the ride. It distributes demand and reduces the overall wait time.

These examples highlight the diverse strategies companies across various industries employ to manage and improve wait times.

The lesson here is clear. Any investment that reduces your wait time will equal happier customers who will ultimately spend more and return time and time again. Don't fall into the trap of the way you've always done things. While new technologies will cause growing pains, customer expectations are always increasing, and you must use every tool in your arsenal to get ahead of the curve to be more responsive.

CHAPTER 10

REAL-TIME RESPONSE

"If watching is all you're going to do,
then you're going to watch your
whole life go by without you."

– Laverne, *The Hunchback of Notre Dame*

Peter Drucker famously said, "what gets measured, gets managed." Allow us to offer a corollary. "What gets measured in real-time, gets managed in real-time." We've talked at length about decreasing attention spans, the imperative to know your customers and the art of reducing wait times. All those concepts come smashing together in this category of examples we call real-time response.

The most magical place on earth happens to know a few things about real-time response. In a *New York Times* article, Disney executives gave a rare look into the park's operations. Deep below Cinderella Castle is the Disney Operational Command Center. Technicians there are watching wait times at every Magic Kingdom attraction and deploying various tactics to alleviate congestion in real time.

> In one corner, employees watch flat-screen televisions that depict various attractions in green, yellow, and red outlines, with the colors representing wait-time gradations. If Pirates of the Caribbean, the ride that sends people on a spirited voyage through the Spanish Main, suddenly blinks from green to yellow, the center might respond by alerting managers to launch more boats. Another option involves dispatching Captain Jack Sparrow or Goofy or one of their pals to the queue to entertain people as they wait. "It's about being nimble and quickly noticing that, 'Hey, let's make sure there is some relief out there for those people,'" said Phil Holmes, vice president of the Magic Kingdom[69].

By monitoring and responding in real time, Disney can manage wait times and provide a less congested experience for customers. Waiting is a fact of life at any theme park; however, Disney's monitoring

69. http://www.nytimes.com/2010/12/28/business/media/28disney.html

and responses enables them to keep guests happy by maximizing capacity, directing flow through the park and providing entertainment in those moments when waiting is inevitable.

KEEP IT SIMPLE

In an era where complexity often overwhelms, Perplexity AI emerges as a breath of fresh air, simplifying our interaction with data and content creation. This brainchild of Andy Konwinski, Aravind Srinivas, Denis Yarats, and Johnny Ho, established in 2022, swiftly soared beyond the capabilities of its forerunner, ChatGPT. It's not just another AI tool; it's a paradigm shift in how we access and use information.

At its core, Perplexity AI is a testament to the advancements in language models, leveraging deep learning to not just churn out content, but to do so with a finesse that captures relevance and accuracy. Whether you're a student, a professional, or just a curious mind, Perplexity AI stands ready to transform vast data into precise, context-rich insights. A key benefit is Perplexity provides verifiable citations so that the user can understand where the data is coming from and thereby evaluate its relevance.

Perplexity AI's adaptability is one of its crown jewels. Whether you're uploading files, working on cloud platforms, or navigating through GBD4, it molds to your needs. This chameleon-like ability makes it an invaluable ally in various arenas - think of it as your personal assistant in education, professional development, or creative endeavors.

While ChatGPT was groundbreaking, Perplexity AI leaps ahead with its real-time data fetching feature. No longer confined to historical data capped in 2021, Perplexity AI delves into the present, offering insights on the latest happenings. It's like having a living, breathing encyclopedia that not only provides information but also cites sources and expands on terms, enriching your understanding and credibility.

What truly sets Perplexity AI apart is its user-friendly design. It's an open door to the world of AI for everyone, no technical expertise required. Quick, accurate responses, high-quality content generation, and a knack for citing sources are just the tip of the iceberg. It's a boon for researchers, creatives, and anyone thirsty for knowledge.

Perplexity AI isn't just an advancement; it's a revolution in artificial intelligence. It breaks down barriers to information access, offering a potent, user-friendly tool for a myriad of applications, from crafting detailed articles to in-depth research. As it continues to evolve, staying ahead in AI innovation, Perplexity AI reaffirms its commitment to making knowledge more accessible, proving itself as a cornerstone in the ever-evolving landscape of AI.

THINK BENEFITS NOT COSTS

Sophisticated monitoring and response might seem like an expensive endeavor. When effective, however, you'll find more benefits than costs.

Enevo has developed a wireless smart waste management system that has transformed the financial, environmental, and social impact of waste. The system utilizes smart sensors placed in trash and recycling bins to measure the level of their contents. This data is then collected and analyzed using a cloud-based application, providing actionable insights to right-size waste management services, validate the services provided, and improve asset utilization.

By implementing Enevo's smart waste management system, customers have reported significant cost reductions. The system has helped cut costs by 50 percent through a reduction in the number of waste collection trucks needed on the road, leading to a 50 percent decrease in costs for Enevo's customers. This reduction in the number of collection trucks not only results in cost savings but also contributes to a decrease in fuel consumption and associated emissions, aligning with environmental sustainability goals.

Furthermore, Enevo's system has been instrumental in improving the efficiency of waste management services. The use of smart sensors to track and monitor the level of waste in bins allows for optimized waste collection routes, reducing unnecessary trips and minimizing the environmental impact of waste collection. This enhanced operational efficiency not only leads to cost savings but also contributes to overall environmental sustainability by reducing fuel consumption and emissions.

Enevo's smart waste management system has proven to deliver tangible benefits that exceed its costs. By providing actionable insights, cutting costs, and improving operational efficiency, the system has made a significant impact on the waste management industry, contributing to a more sustainable and cost-effective approach to waste collection and disposal.

PROVIDE TRANSPARENCY TO CUSTOMERS

Having real-time data also opens up a world of customer-facing status updates to keep them in-the-know, which creates an overall feeling of empowerment and control. Australian telecommunications provider Telstra does just that. In March 2014 they announced a bold plan to provide real-time information to customers. The customer-facing enhancements include a real-time view of billing and payments, instead of its current solution which is delayed by 24 hours. In November 2023 real time data became instantaneous.

In addition to billing and payments, Telstra is using geolocation technology to track its technicians more efficiently in the field, deploy them intelligently and keep customers abreast of when a technician will arrive. The result is less time waiting at home for a technician to arrive and a more efficient delivery model, which utilizes technician time most effectively, creating a win for the company and the customer[70].

70. http://www.itnews.com.au/news/telstra-to-offer-real-time-billing-tracking-of-technicians-373811

Uganda's leading online food delivery marketplace, Hellofood, knows this lesson as well. They have a smartphone-based solution which tracks exactly where delivery drivers, commonly called boda bodas, are located and can dispatch them most efficiently. Their solution allows them to accurately estimate the amount of time it will take for the driver to pick up and deliver the food, allowing them to deliver an accurate delivery estimate and real-time status updates.

Ron Kawamara, Managing Director HelloFood Uganda, said, "Customer experience is of utmost importance to us, and as such we continue to optimize and incorporate the latest technology to improve on it. And the obvious place to start is the delivery and operations systems. The software uses GPS to track the location of the riders and also furnish them with the exact orders placed by the customers which is better than calling them."

"The software also helps streamline the dispatch process for the riders and thus operation system is now more efficient. This technology also helps the company maintain a level of transparency with customers who may be anxiously waiting on their meals and would like to inquire how far along it will be before their meal arrives either to their office or home; as we are now able to accurately communicate with customers about their arrival of their meal, with accurate follow-ups," Ron added[71].

Besides Hellofood, several other companies use real-time tracking for food deliveries. The trend started well over a decade has become mainstream for food deliver companies. Some of these companies include GPS Leaders, which offers GPS tracking for food delivery companies, and Shipday, a dispatch and delivery tracking software designed to help restaurants, courier services, and other small businesses plan orders. Additionally, Onfleet, Tookan, and

---

71. http://pctechmag.com/2015/09/hellofood-introduces-software-to-track-your-food-delivery-in-realtime/

HungerRush 360 are also among the companies that provide food delivery software with live tracking features. Another example is Locate2u, which offers food delivery software that provides real-time tracking and live location sharing for optimized routes. These companies utilize real-time tracking to improve delivery efficiency and customer experience.

Consumers have come to expect the idea of food delivery, advanced dramatically because of the pandemic, and as such, companies such as Uber Eats, Instacart, Grubhub, and DoorDash have flourished. And the key is not the outcome – food delivery – but the ability to show in real time the process from order to delivery.

BEYOND HOME SERVICE AND DELIVERIES

Real-time response technology also helps financial organizations provide better service and look after the bottom line. Take Pioneer Federal Credit Union for example. The collections team at Pioneer used to receive reports on delinquent loans only on a monthly basis. This meant that the first action to help members resolve potential financial difficulties might be delayed by as much as four weeks. The organization now delivers a daily report on loans that are already or soon to be delinquent.

The real-time reporting, enabled by IBM Cognos, has driven impressive results for Pioneer. "The insight delivered by Cognos TM1 has helped us to bring delinquency to a historic low – I can't overemphasize the magnitude of that achievement," said CFO Dan McGowan. "Our delinquency stood at 147 basis points, and no-one in Pioneer could ever remember it being below 100 basis points. [One year later] we were down to 37 basis points. Our COO attributes this to that daily visibility provided by Cognos TM1, which enables the collections group to be proactive in helping our members meet their commitments. Early intervention means fewer loans charged off and enables us to be more member-friendly than a lot of other institutions."

Wells Fargo takes a different tact with their Your Mortgage Dashboard tool. In the words of Thomas Smith:

> Wells Fargo's loan tracker is similar to Dominos popular pizza tracker tool. It enables customers to quickly view the status of their mortgage from a variety of devices and provides reminders on next steps or outstanding loan requirements. This tool is great for customers because it gives them more insight into a typically long and confusing process, while also reducing simple customer services calls checking on the loan status. Your Mortgage Dashboard seems like a relatively simple solution for what I would suspect was a significant problem for customers. It also is a great way to keep customers engaged throughout a long product application process. There are a number of financial services products that fall into this category and could benefit from this kind of solution.

Whether you're reaching out to customers who might be in financial trouble or helping them understand where they are in a complicated process, service-based companies have significant opportunities to stand out from the competition, just as Pioneer and Wells Fargo have done.

FIVE ADDITIONAL EXAMPLES

Here are five additional examples of companies excelling in real-time responsiveness:

1. **American Airlines' Real-Time Baggage Tracking:** American enhances passenger peace of mind by offering the ability to track checked baggage in real-time via its app. This feature keeps passengers informed about the status of their bags throughout their journey.

2. **Starbucks' Mobile Order & Pay:** Through its Mobile Order & Pay feature, Starbucks allows customers to order and pay ahead via the app. The app provides real-time updates on order preparation and pickup times. This significantly streamlines the pickup process and improves overall customer satisfaction.
3. **Dominos' Pizza Tracker:** Dominos' Pizza Tracker keeps customers informed with real-time updates on their pizza order, from the moment it's being prepared to the second it's out for delivery. This transparency enhances the ordering experience by involving customers in the process and managing their expectations in real-time.
4. **Uber's Live Trip Sharing Feature:** Uber's live trip sharing feature allows riders to share their real-time trip status with friends or family, including estimated time of arrival and live location tracking.
5. **Amazon's Prime Now:** Amazon Prime Now offers customers real-time tracking of their orders with the ability to see the exact location of the delivery driver on a map when they are close by. This service provides a high level of transparency and convenience, particularly for urgent purchases.

These examples demonstrate how leveraging real-time data and updates can significantly enhance customer service and experience.

Taken together, the message of real-time response is clear—the more you know and the faster you can respond allows you to create differentiation. Sometimes it's surprising and delighting a customer with speed. Other times it's identifying and correcting an issue before a customer even knows about it. And when necessary, it's responding to correct a bad situation before it gets worse.

There's no doubt the technical hurdles are steep, but don't fall into the trap of sameness. Tackle the hurdles and choose to do more for your customers in real-time.

PART IV:

THE THIRD R

CHAPTER 11

READINESS

"Don't wait until you are ready to take action.
Instead, take action to be ready."

– Jensen Siaw

In 2008 Disney set off on its Next Generation Experience project with the intent to remove all friction from the Walt Disney World experience. The problem, on the whole, was that families spent more time planning and coordinating their trip and less time enjoying it. Less time making memories means less mindshare for Disney and a lesser chance of a return visitor.

WIRED published a lengthy overview[72] of the project and while we can't go into every detail here, the result of the project was Disney's MagicBand. The futuristic wristband, a quintessential example of form and function, allows short- and long-range sensors around the park to identify you. The result? Magic.

With just this identifier, Disney has eliminated a handful of other items to carry and handle. Park tickets? Ready on the MagicBand. Photo Passes? Tagged to your account automatically. Room keys? Replaced by the band. Fast Passes? Schedule them online and check in with your MagicBand. Cash? Just swipe your MagicBand to bill a credit card on file.

With everything coordinated and synchronized with your Disney account and its smartphone app, you're able to experience the park in a new way. Planning Fast Passes in advance allows you to schedule your must-do attractions and allow spontaneity to guide the rest of your trip.

Beyond the surface, Disney can use the bands and sensors to create magical experiences. Their signature restaurant Be Our Guest allows you to order in advance and sit down at any table. Magically your food is delivered, without your server even asking your name. It's all powered by the MagicBand and a series of sensors.

72. http://www.wired.com/2015/03/disney-magicband/

The MagicBand is truly a platform project with many future implications from enabling characters to know your name without asking or detecting a long wait and rewarding you with another FastPass or a free ice cream coupon. The possibilities are simply endless.

As would be expected Disney's MagicBand has undergone improvements and is now called MagicBand+. This latest iteration offers several added benefits. It retains the original MagicBand's features, such as storing park tickets, Lightning Lane reservations, and enabling purchases. However, the MagicBand+ introduces new functionalities, including a redesigned look with a watch-like silhouette and haptic vibrations for enhanced sensory experiences at the parks. It also incorporates interactive park experiences through motion recognition, allowing guests to engage with various elements in the park with just a wave of their hand. Additionally, the MagicBand+ is rechargeable, unlike the original MagicBand, and is compatible at both Walt Disney World and Disneyland. While the MagicBand+ comes at a higher cost, it offers a range of new and immersive features that enhance the overall Disney Park experience.

Your lesson here is one of planning and preparation. Creating magic requires an invisible effort of planning and design. When Disney set out on this project, they weren't responding to problems one by one; rather, they set out to build a system that removed friction from the Disney experience. This is what we can readiness.

We've said countless times that expectations are increasing, and you must get (and stay) ahead of the curve. Readiness is all about putting systems and processes in place now, so you'll be able to surprise and delight when the opportunity arises. It's exactly what Disney has done, and you can do so as well.

CHAPTER 12

FRICTIONLESS COMMERCE

"Computers have enabled people to make more mistakes faster than almost any invention in history, with the possible exception of tequila and handguns."

– Mitch Ratcliffe

Helping a customer do business with you without friction is almost a sure way to increase revenues. When you're a staple of quick service like Starbucks, time is money. Waiting for customers to find payment methods can make the difference between a reasonable line and one that causes customers to turn the other way and leave.

Starbucks has been a key player in popularizing mobile payments and loyalty programs. As of 2022, customers can use various payment methods at Starbucks, including Apple Pay, Google Pay, PayPal, Venmo, and credit/debit cards. The Starbucks app for iPhone and Android allows customers to pay, track rewards, and order ahead at participating locations. The Starbucks app is the second most used mobile payment app for point-of-sale transactions in the US, right after Apple Pay, and is used by over 30 million Americans. Starbucks' loyalty program has been successful in encouraging mobile ordering, with about 7 million orders placed through mobile devices in US cafes in February, and over 12 million Starbucks Rewards members as of 2021. The company's focus on the mobile experience has been a core strategy, with a significant portion of its transactions now coming from mobile payments and orders.

Starbucks is clearly winning with mobile payments and they're not the only ones fighting friction. Traditional payments providers, including American Express, PayPal, **Chase Paymentech**, Stripe and Square all offer mobile payment solutions for businesses, including the ability to accept payments via mobile devices.

The Tampa Bay Lightning took the concept of frictionless payments a step further and turned it into a value add for their best fans, season ticket holders. Along with their tickets, the Lightning delivered an exclusive jersey to each fan. Many fans would be in awe over an exclusive product not available to others. This jersey was even more special. As presented at the National Sports Forum:

To create added value for season ticket holders, the Tampa Bay Lightning created a revolutionary jersey to give to all full-time season ticket holders. To give the jerseys a distinct look and subsequently unite fans at home games, the Lightning adorned the jerseys with a "season ticket member" patch and allowed fans to customize the jerseys with their own name and number. Due to Tampa Bay's desirable climate and large number of travelers, it's not uncommon for out-of-town jerseys to outnumber Lightning jerseys in the stands. Giving away these special jerseys to the season ticket holders helped to increase team colors at Lightning games at the Times Forum.

In addition to making them feel like a member of the team, the Lightning wanted to incentivize their most loyal fans to purchase while in stadium. "Season ticket members commit 41 nights a year to be in our building, so we felt they deserved benefits that no one else could receive," said Lott. "We noticed that season ticket holders weren't buying food in-stadium, they were opting to eat out and we wanted to incentivize them to try out our new offerings." To get them to buy in-stadium, the team embedded RFID chips into the sleeve of each jersey. The chips can be scanned throughout the stadium for an exclusive discount: 25 percent off concessions and 35 percent off merchandise. Lott describes the creation of the promotion, "It started as a joke, someone mentioned putting a chip in our fans, but then we realized we could make this a reality." The Lightning partnered with Quest and Radiant Systems to create the RFID chips. They created a durable chip about the size of a quarter that can be dry-cleaned, ironed, and washed 10,000 times.

> Confident in the RFID chip and its ability to work, the team installed an RFID reader at every Point-of-Service station in the Times Forum building. The reader scans the chip, automatically recognizes who the season ticket member is and gives them their discount. The transaction information is recorded, and the team can see buying patterns and where season ticket members are buying from. This information allows the team to better understand their consumers to create smarter, more effective promotions. "The beauty of this system is that it is segmented out based on the area of the arena that a season ticket holder sits in. It gives us the ability to do promotions based on segmented areas throughout the arena" adds Lott[73].

Reports shows the RFID-enabled jerseys were a huge hit. The Lightning saw revenue increase in all categories. Season ticket sales are up nearly 100 percent with concession and merchandise sales up 25 percent.

We love this example because it combines the idea of rewarding your best customers and using technology to reduce friction. It's a win-win.

Amazon is leveraging technology to enhance the shopping experience through its Dash Replacement and Auto Reorder features. The Dash Replacement program, which includes the Dash Button and Smart Reorders, enables customers to automatically reorder common household items with a simple touch or voice command. The Dash Button, introduced in 2015, allows users to reorder products instantly with a physical button press, while Smart Reorders, integrated with Alexa, automatically purchases replacement items when supplies are running low. Furthermore, the Auto Reorder

73. http://sports-forum.com/news/new_archive/index.html?article_id=453

feature, part of Amazon Dash Replenishment, enables customers to manage their settings and toggle the auto reorder switch on or off for supported devices, ensuring a seamless and convenient shopping experience.

By integrating these technologies, Amazon aims to simplify the process of replenishing everyday essentials for its customers. The combination of physical and virtual one-touch reordering, along with smart reordering capabilities, demonstrates Amazon's commitment to leveraging technology to make shopping more convenient and efficient. These features not only streamline the shopping experience but also contribute to the seamless integration of connected devices into the household, ultimately enhancing customer satisfaction and loyalty.

On March 1, 2019, Amazon discontinued the series, claiming that it was made unnecessary due to automatic reordering and product subscriptions. Additionally, Amazon claimed that voice-activated shopping on Alexa products would succeed the buttons. As of this writing, the success of voice-activated shopping with Alexa is yet to be proven widely successful.

Amazon's Prime Air drone delivery service will deliver to customers in three U.S. locations as well as cities in Italy and the United Kingdom by the end of 2024. The newest Prime Air drone, the MK30, has been unveiled as part of the latest expansion to the Prime Air program, and it's unlike any other drone being used for package delivery.

NOT JUST TRANSACTIONS

Orders and payments are just two ways to reduce friction. Countless other companies are doing new and interesting things to eliminate friction for their customers. In fact, one insurance startup named Oscar, differentiates on reducing friction.

Oscar policyholders have the best of both worlds. Oscar licenses its provider network from an established insurer and also contracts out claims processing. Where they stand out, however, is providing a technology-enabled experience that puts patients in control.

Oscar features benefits like the ability to call a doctor for a consultation, an online price comparison database and a Google-esque search box that allows policyholders to type in an ailment and see a list of possible solutions. All in all, these benefits accomplish two things. They help provide care to the policyholder as efficiently as possible, providing them a better experience. The business benefit is avoiding costly emergency room trips or guiding policyholders to less expensive options for the same procedure.

Under all its plans, Oscar allows its members unlimited free calls with physicians. The doctors are supposed to call back within an hour, but on average, calls are returned within seven minutes of being placed, founder Kevin Nazemi told *The New York Times.*

One individual who called had taken a tumble in the subway and wondered if his arm was broken, recall Oscar's executives. Over the phone, the doctor took the patient through a range of motions and determined that the arm wasn't broken. He advised putting an ice pack on it and seeing the primary care physician the next day. That saved the patient an expensive trip to the emergency room, Oscar's founders said[74].

SUBSCRIPTIONS

Another popular way companies are reducing friction is through subscription programs. These programs often provide convenience

74. http://www.nytimes.com/2014/03/29/business/start-up-health-insurer-finds-foothold-in-new-york.html

to customers at a lower price point than they'd pay purchasing frequently used items one at a time.

Our favorite example is Dollar Shave Club. The star of its viral YouTube launch video that went viral, CEO Michael Dubin told *The Wall Street Journal*, "We have 2 million members that get a shipment every month or every other month," Dubin said. "Men's grooming is exploding and we think we have a role to play.[75]"

Packages range from $3.00 to $9.00 per month. Each month, the company emails subscribers to confirm their order and provide a one-click option to order add-ons like shave gel and aftershave lotion. This added convenience helps to drive sales of ancillary products, but also does a great job of keeping customers informed and in control.

The company also takes special care to celebrate its customers, profiling them in its monthly newsletter, *The Bathroom Minutes,*[76] which is incidentally designed for reading in the bathroom.

So successful, Dollar Shave Club was acquired by Unilever for $1 billion. Subsequently, in 2023, Unilever sold control of Dollar Shave Club to a private equity group, but the terms of the sale were not disclosed. Obviously, the idea, so well received, has shown the power of a well-placed subscription service.

Active as an independent business today, the business model of Harry's razors is based on direct-to-consumer distribution, vertically integrated manufacturing, and a subscription service. This model allows the company to cut out the middleman, control the design and quality of its products, and create a recurring revenue stream through subscriptions.

75. http://blogs.wsj.com/digits/2015/06/21/dollar-shave-club-valued-at-615-million/
76. https://contently.com/strategist/2015/05/26/dollar-shave-club-wants-to-keep-you-reading-on-the-toilet/

By selling a limited line of razors, owning its manufacturing facility, and offering a subscription service, Harry's has been able to disrupt the razor industry and achieve significant growth, reaching a valuation of $1.7 billion in 2021. Additionally, the company has adapted to changing market dynamics by launching in-store partnerships with retailers and expanding into new product categories.

North Carolina-based FilterEasy has done just the same for home air filters. FilterEasy is an air filter service that's smart and simple. You sign up once, then they deliver your specified filters to your door on your schedule every time you need them. You pay when your filters are shipped, and you can change your preferences at any time.

In addition to saving time and money, Filters arrive at your door just when they're supposed to—whether or not you remembered they were on their way. This provides a gentle reminder to change your filters regularly, resulting in a healthier home. Another win-win for the company and its customers[77].

HANDLING COMPLEXITY

An easy conclusion about reducing friction is that it's only for simple transactions. Oakland-based Sungevity shows us otherwise. They've mastered the art of creating a seamless, frictionless experience around a very complicated product, residential solar panels. David C. Edelman and Marc Singer covered the company in a recent *Harvard Business Review* article:

> Oakland-based Sungevity, competes on its ability to shape the customer journey. At first glance, Sungevity looks like a typical residential solar panel provider. But closer inspection reveals that the company's business is to manage the end-to-end process of sales and

77. https://www.filtereasy.com/

custom installation, coordinating the work of an ecosystem of companies that supply, finance, install, and service the panels. Sungevity's "product" is a seamless, personalized digital customer journey, based on innovative management of data about the solar potential of each home or business. Sungevity makes the journey so compelling that once customers encounter it, many never even consider competitors.

The process began when he received a mailing with the message "Open this to find out how much your family can save on energy costs with solar panels." The letter within contained a unique URL that led to a Google Earth image of the customers house with solar panels superimposed on the roof. The next click led to a page with custom calculations of energy savings, developed from Sungevity's estimates of the family's energy use, the roof angle, the presence of nearby trees, and the energy-generation potential of the 23 panels the company expected the roof to hold.

Another click connected the customer through his desktop to a live sales rep looking at the same pages the customer was. The rep expertly answered his questions and instantly sent the customer links to videos that explained the installation process and the economics of leasing versus buying. Two days later, Sungevity emailed the customer with the names and numbers of nearby homeowners who used its system and had agreed to serve as references. After checking these references, the customer then returned to Sungevity's site, where a single click connected him to a rep who knew precisely where he was on the journey and had a tailored lease ready for him. The rep emailed it and walked the customer through it, and then it was

> e-signed. When the customer next visited the website, the landing page had changed to track the progress of the permitting and installation, with fresh alerts arriving as the process proceeded. Now, as a Sungevity customer, the customer receives regular reports on his panels' energy generation and the resulting savings, along with tips on ways to conserve energy, based on his household's characteristics.
>
> Starting with its initial outreach and continuing to the installation and ongoing management of the panels for the customer, Sungevity customized and automated each step of the journey, making it so simple—and so compelling—for them to move from one step to the next that he never actively considered alternative providers. In essence, the company reconfigured the classic model of the consumer decision journey, immediately paring the consideration set to one brand, streamlining the evaluation phase, and delivering the customer directly into a "loyalty loop," where they remain in a monogamous and open-ended engagement with the firm. Sungevity's journey strategy is working. Sales have doubled in the past year to more than $65 million, exceeding growth targets and making Sungevity the fastest-growing player in the residential solar business[78].

Sungevity realized their success depended upon a seamless customer journey and their investments are paying off. In fact, all of the companies in this chapter are seeing benefits from simply making it easier for the customer to do business with them. They created a successful model that resulted in their acquisition in 2020.

78. https://hbr.org/2015/11/competing-on-customer-journeys

Pineapple Energy is a U.S.-based operator and consolidator of residential solar, battery storage, and grid services solutions. The company has been involved in a series of acquisitions and mergers, including the acquisition of Sungevity's assets in 2020. Pineapple Energy resumed activity under the Sungevity name in California only. Additionally, Pineapple Energy merged with Communications Systems Inc. and changed its name to Pineapple Holdings Inc. The company's focus is on the rapidly growing home solar industry, with a particular emphasis on residential distributed energy resources and virtual power plants.

FOUR ADDITIONAL EXAMPLES

Here are four additional examples of companies that have successfully reduced friction from the customer experience:

1. **Amazon Go:** Amazon Go stores eliminate the traditional checkout process through their Just Walk Out Technology. Customers enter the store using the Amazon Go app, pick up the items they need, and leave without having to check out. The store uses sensors and AI to track what items are taken. Just leave and the app automatically charges your Amazon account.
2. **Apple Pay & Face ID:** Apple enhances the payment process with Apple Pay and Face ID, allowing users to make secure purchases in stores, in apps, and on the web without having to manually enter payment information. The use of biometric authentication not only speeds up transactions, but also adds a layer of security.
3. **Shopify's Shop Pay:** Shopify offers Shop Pay, a fast and secure payment option that stores customers' shipping and billing information so that it can be quickly accessed for future purchases across Shopify's vast network. This feature significantly speeds up the checkout process for returning

customers. It not only makes online shopping quicker, it also reduces cart abandonment.

4. **Nike's SNKRS App:** Nike's SNKRS app provides a streamlined way for sneaker enthusiasts to purchase new releases without having to navigate through the traditional online shopping hurdles. By offering exclusive access, detailed product information, and a seamless checkout process, the SNKRS app caters to its community with a tailored, frictionless buying experience.

These examples showcase how various industries are implementing innovative solutions to minimize customer effort, speed up transactions, and enhance overall satisfaction by reducing friction in the commerce experience.

The overarching lesson is to think of reducing friction note as an expense but as an investment. By deploying smart programs with your customers' needs and desires in mind, you'll certainly create both profits and prophets for your business.

CHAPTER 13

LOCATION-AWARE CONVENIENCE

"A vast technology has been developed to prevent, reduce, or terminate exhausting labor and physical damage. It is now dedicated to the production of the most trivial conveniences and comfort."

– B.F. Skinner

Tracking customers by location used to seem a bit invasive, but with new possibilities of providing better service or incentives, customers are warming up to the idea of sharing more with companies. Companies are using location awareness to accomplish all sorts of enhancements from knowing when a customer has arrived to helping them navigate a complicated building.

While we won't cover them much here, we also can't ignore the benefits provided from an analytics standpoint. Companies who can collect location information from customers can learn more about those customers and provide them with a customized experience.

Nearly every brick-and-mortar retailer is working on some sort of location-aware experience. While many of these provide special offers or deliver marketing messages, we've homed in on a few that specifically benefit the customer.

One that caught our attention was Virgin Atlantic's use of beacons in Heathrow airport to help guide travelers through the travel experience. From the Virgin Atlantic blog:

Upper Class passengers at London Heathrow will become some of the first people to experience a new technology designed to give personal notifications to their phones via Apple's iBeacon technology, which uses Bluetooth Low Energy and geofencing to provide a connected service through their airport experience.

Smartphones, with Apple Passbook downloaded, in range of a device can receive communications from the beacon related to their location in the airport. For example, an Upper Class passenger approaching the private security channel can receive a notification for their phone to open their electronic boarding pass ready to be scanned by security. In the main area of the airport, passengers using the service will receive special partner offers, such as zero percent commission as they pass the MoneyCorp currency exchange booth.

"At this stage, we've only skimmed the surface on the advantages that this technology can offer our customers," said Reuben Arnold, our Brand & Customer Engagement Director. "In the future, we expect it to further personalize our passenger's journey, such as notifying an Upper Class passenger that there is an open appointment in the Clubhouse spa or introducing the crew members on their flight as they board the aircraft. Our passengers are increasingly more connected as they travel and by using advanced technology, we can improve customer experience and continue to put innovation at the heart of the flying experience."

"The way we're looking at iBeacon is probably not quite the standard use of it," said James Shanahan, our Head of Development for eBusiness. "We're talking in terms of the surprise and delight factor, and so while there may be the potential for future discounts and offers, we're actually looking at how we can provide a greater customer experience."

This closing quote highlights exactly why we chose to feature Virgin Atlantic. Their customer-first approach to technologies often results in a far better experience for customers. Virgin's foray into the use of this technology has been followed by British Airways who implemented an iBeacon initiative at Terminals 3 and 5 at Heathrow Airport, Japan Airlines who installed beacons at Tokyo Haneda airport and provided staff with smartwatches to support the initiative and American Airlines that undertook a significant deployment of iBeacons at Dallas/Fort Worth International Airport.

Tesla has earned a similar reputation. One of our favorite examples from Tesla, and there are many, is how they roll out new features to existing customers. Most software updates for cars are conducted at a dealership and tend to be rather expensive. Tesla takes a different approach and rolls out new software automatically to all existing customers. In fact, they often place hardware in vehicles before the software is ready. This means that existing customers won't miss out on new features, which is a huge win for customers.

Chief Executive Elon Musk said in a press conference, "We view it in the same way as updating your phone or your laptop," he said. "It is a fundamental paradigm shift from the way that cars have been done in the past."

Early updates added a new set of features which analyzes a driver's route, road conditions, and even topography. The software tells the driver if the car is going out of range of a charging location. This helps alleviate consumer fears of running out of charge without the ability to reach a charging station in time, a concept called "range anxiety[79]."

The added benefit of knowing you won't go out of range in your Tesla is a significant benefit to customers. No auto manufacturer wants to be associated with a roadside issue and Tesla is avoiding just that with these new features.

Of course, while Tesla pioneered the concept and dominates the EV market, brands such as Mercedes, Lucid, Audi, Volvo, and Land Rover/Jaguar all use over the air updates. And those updates aren't limited to EV's only as many hybrid models now find value in the adoption of technology.

Another way companies are using location technology is to provide useful updates to nearby consumers. Real estate app Trulia has taken this concept to the housing market. Especially useful in competitive housing markets, the app can alert you to when a new listing is added nearby within the last 24 hours.

79. http://www.wsj.com/articles/tesla-to-add-car-safety-features-via-downloads-1426785366

You have control of the alert frequency and the radius of nearby alerts. This eliminates the manual work of searching for nearby listings from time to time. The increased efficiency could be the difference in finding your perfect home or losing out to quicker competitors[80].

HELP NOT HYPE

Remember the movie "Minority Report" where Tom Cruise's face was recognized, and it created personalized ad for his attention. Well, that's not far from reality.

The FaceMe Platform, developed by CyberLink, is a facial recognition technology that accurately identifies characteristics such as age, gender, and mood, enabling customized ads and targeted marketing for smart retail. It verifies identity by comparing facial templates, supports liveness detection with 2D cameras, and allows the management of specific facial images and grouping people according to specific criteria.

An example of the FaceMe Platform in use today is its application in smart retail. The platform can be used to identify the characteristics of individuals entering a retail store, such as their age and gender, and then deliver customized ads and marketing messages to them based on this information. This allows for a more personalized and targeted approach to engaging with customers.

The use of facial recognition technology for targeted advertising, as exemplified by the FaceMe Platform, does bear some resemblance to the concept shown in the movie *Minority Report*. In the movie, personalized ads are delivered to individuals as they move through public spaces, based on their retinal scans. While the current use of facial recognition technology in advertising is not as advanced

80. http://venturebeat.com/2014/05/15/trulias-apps-can-now-alert-you-of-new-home-listings-nearby/

as depicted in the movie, it does share the fundamental idea of delivering highly targeted ads to individuals based on their specific characteristics and behavior.

The FaceMe Platform and similar technologies demonstrate the increasing integration of facial recognition into marketing and customer engagement strategies, raising important considerations around privacy, data security, and the ethical use of such technologies in commercial settings.

FOUR ADDITIONAL EXAMPLES

Here are four additional examples of companies utilizing location-aware technology to enhance convenience and customer service:

1. **Google Maps Real-Time Sharing:** Google Maps allows users to share their real-time location with friends or family, making it easier to coordinate meetups or ensure the safety of loved ones. This feature adds a layer of peace of mind for users.
2. **Target's In-Store Product Location Services:** Target uses its app to provide customers with a detailed map of the store, showing the exact location of products on their shopping list. This minimizes the time customers spend searching for items and makes shopping more efficient.
3. **Find My iPhone:** Apple's Find My iPhone feature uses location technology to help users locate their lost or stolen devices. This service not only provides peace of mind but also a practical solution to a common problem. It leverages location awareness for your device to enhance customer satisfaction.
4. **Walgreens Prescription Refill Reminders and Store Locator:** Walgreens uses location-based notifications to remind customers to refill prescriptions when they are near a

store. The app also helps users find the nearest Walgreens location, making it convenient to pick up prescriptions.

These examples illustrate the diverse applications of location-aware technology across various industries. By leveraging real-time location data, companies can offer enhanced convenience and personalize the customer experience.

Location-based marketing is all the rage with technologies designed to push messages to consumers. We believe companies should look past broadcast-based thinking and consider how these technologies can benefit the customer. Take a cue from the companies featured in this chapter to up your game.

CHAPTER 14

TRUST AND SAFETY

"The only real security that a man will have in this world is a reserve of knowledge, experience, and ability."

– Henry Ford

When compiling examples for Blue Goldfish 2.0 this trend took us by surprise. Using technology to help keep customers safe certainly seems like common sense, however we all know that common sense is rarely common practice.

Dutch train operators Prorail and NS caught the attention of Trend-Watching.com as one example of a company using info-sense to improve customer experience. The duo has rolled out LED displays that provide real-time service and safety information to passengers. The 600-foot-long (182 meter) displays are akin to what you might envision circling the balcony of your favorite sports arena. On the platform these displays indicated how crowded each carriage car is, where the doors will open, which ones can accommodate bikes and where you can find the quietest car.

Infrared sensors on the train cars can determine how full each car is and communicate that information to the platform displays in advance, allowing passengers to select which car they want to use and gather in the appropriate position.

Speaking about the new displays, Caroline Hummels of the Eindhoven University of Technology said, "the passenger information displays beauty and power lie in its intuitiveness. It's such a simple, natural solution, one wonders why it took so long."

We believe that's a good measuring stick for the examples in this chapter and have compiled examples of intuitive and seamless safety measures companies are including in their products and services.

TWO-WAY REVIEWS

A common trend of features which enable a safer experience is two-way reviews. Famously used by eBay in the early days, the notion of reviewing both the provider and the customer have become a staple of modern sharing services.

We've all seen this customer. He or she is rude to the staff, complains incessantly, and becomes a general nuisance. Up until recently there was little or no recourse if you were a seller or service provider. You simply had to take it. These vampires drained employee morale and took the joy out of the equation for other customers.

Enter Uber and Airbnb, both of which have the changed the rules. Uber allows drivers to rate customers and Airbnb allows hosts to leave reviews of guests. It is now transparent on both sides of the register. Technology has leveled the playing field. Your history as a buyer follows you from one transaction to the next. Poor behavior may keep you from getting a ride or staying at the apartment of your choice.

Will this change the behavior of customers? Are customers now cognizant of their behaviors enough to improve? Perhaps so. Here's the story of an Uber customer in *The Chive,* in his own words:

> Tonight, I got an Uber ride. After the driver was on their way, they sent me a text message telling me that they're deaf and to keep that in mind. He picked me and headed home. On the way, I googled how to say, 'Thank you very much' in sign language. When we got to my house and he pulled over, he turned back to look at me and I signed, 'Thank you very much' and he started crying. He shook my hand firmly with two hands, wiped the tear from his eyes and signed back 'Thank you' several times. Something simple that I did meant a lot to someone else. It was a good day.

It is doubtful that this passenger was motivated by a mere Uber rating. But it is reasonable to think that the knowledge of being observed will impact behavior. Researchers label this the Hawthorne effect. It is a type of reactivity in which individuals modify or improve an aspect of their behavior in response to their awareness of being observed.

DOES POLITENESS PAY?

If politeness doesn't pay, at least in one place rudeness will cost you extra. According to Trendwatching.com, there is a café in France that has implemented a pricing policy based on patron's politeness. Patrons who greeted the barista at La Petite Syrah in Nice and used "please" were charged 1.40 euros/$2 for a cup of coffee; those who failed to use any pleasantries were penalized with a 7 euro/$10 price. Caveat emptor: the prices (with greeting) were clearly displayed on a board inside the café.

Bad customers beware. Companies who value employees are not going to put up with a pattern of misbehavior. According to James Heskett and Earl Sasser, companies such as ING Direct (now Capital One) stand up to protect their employees. The bank has an operating strategy based on a strong, effective culture is selective of new customers. ING Direct had a practice of firing thousands of customers every month. This strategy is especially important when customers abuse policies or make unreasonable demands. Zappos employs a similar strategy of firing customers who abuse policies or employees. It's one of their core seven tenets of achieving exceptional customer service, "Fire customers who are insatiable or abuse your employees."

With Uber and Airbnb, bad behavior will get you banned from the service altogether. Uber has confirmed that riders have been suspended or barred from using the app because of such behavior.

The lesson here is that two-way reviews often protect the customer but can have great benefits for the company as well. As you consider your own programs, we suggest you consider how your review systems will empower employees, help modify the behavior of customers and help you respond when it's time to part ways with a customer.

KEEP IT SIMPLE

Of course, Uber and Airbnb have complex rating systems because they're at the core of their business models. As you consider your own solutions, know that it's often a simple step that helps the customer most.

One great example is Safelite's confirmation emails. When you receive a confirmation email from Safelite, it includes a photo of the technician who will be delivering your service. Have you ever scheduled an appointment with a furnace repairman or lawn service provider, and then thought, "Wait, do I even know or trust this person being sent to my home?" If so, you're not alone.

To set the customer's mind at ease about the mobile technician coming to fix their car, Safelite developed the Technician Profile Email containing the name, brief bio, credentials, and photograph of the technician who will perform the service. The email, which is sent the night before or the morning of the service, also provides the estimated time of service and location as a confirmation.

According to Safelite, the program has been extremely well received. In fact, 85 percent said they felt more comfortable with the technician coming to their home once they receive the email.

As one blogger recently wrote, "Some businesses hire independent contractors to provide services so there is no logo vehicle provided and no requirement the workers wear uniforms. This has been the case for the cable repair people who have been to our home over the last few years. Can you use the Safelite practice to inspire confidence and provide added security for your customers?"

While simple, one important element of Safelite's email is putting a face to the brand.

USING BIG DATA TO SPOT TRENDS

Another apparent theme in our examples is how companies are using big data to detect nefarious activity and respond accordingly.

We're accustomed to crime visualizations from companies like Zillow and Trulia, however these tend to show you where crime tends to happen and don't necessarily predict crime. The Los Angeles and Santa Cruz police department and PredPol, however, are on the case.

PredPol has implemented an algorithm previously used to predict earthquakes and has tweaked it to accept crime data. According to Tech Target, the software can predict where crimes are likely to occur down to 500 square feet. In LA, there's been a 33 percent reduction in burglaries and 21 percent reduction in violent crimes in areas where the software is being used.

However, PredPol, having undergone a reidentification and is now known as Geolitica. This rebranding was part of a larger trend in the predictive policing industry, as companies sought to distance themselves from the negative connotations associated with predictive policing. The name change was seen as an attempt to provide a new image of "data-driven community policing" and to address the criticism of predictive policing. The rebranding was also linked to the introduction of new products that promoted police reform and community engagement. The decision to change the name was made amid public outcry and growing concerns about the use of predictive policing technologies.

Its success has been a topic of debate. An analysis by The Markup found that Geolitica's crime prediction algorithm had a success rate of less than 1 percent in Plainfield, New Jersey. The software's effectiveness has been questioned, with critics arguing that the model is flawed and too simple to effectively predict crime. While there have been reports of success, such as the Alhambra Police Department's positive experience with the software, the overall success of predictive policing, including Geolitica, is still a matter of ongoing discussion and evaluation.

Similarly, Union Pacific has deployed a series of visual and temperature sensors to its tracks. Feeding the data into a predictive analytics engine the company can predict potential derailments before they would likely occur. So far, the sensors have reduced derailments by 75 percent.

BMW Group undertakes a similar process for its vehicles. The group has started to use data gathered from dealers across the world, to spot and fix vulnerabilities. Combined with data from prototype vehicles, which reportedly capture an average of 15,000 data points at any one time, the manufacturer can detect faults before new cars go into production.

While we're not privy to the specifics, the obvious results are certainly attractive. Fewer defects mean fewer recalls and issues, providing a safer experience and potentially less money and time repairing the vehicle later. In both cases there's a win-win, where the company saves money on potential recalls and lawsuits while the customer has a safer vehicle that will likely need fewer repairs[81].

FIGHTING FRAUD

Just as companies are using big data to create safer experiences, others are using similar techniques to fight fraud. Santam, a South African auto insurer has implemented a system which uses predictive analytics to classify incoming claims into categories and, instead of reviewing each manually, takes a risk-based approach in assigning the appropriate team to handle the claim. According to an IBM case study:

81. http://dataconomy.com/5-use-cases-of-big-data-improving-customer-satisfaction/

Santam began working with IBM Business Partner SPSS South Africa to design and implement a claims segmentation solution based on IBM SPSS predictive analytics software. Santam chose IBM and SPSS software because of its range of functionality, flexibility, and its ability to integrate with an existing system. The company's core claims management system resided on a mainframe platform that still met the company's needs. The SPSS solution could be embedded into that system. The SPSS analytics software could also integrate different kinds of rules from across the Santam infrastructure, including process rules from company's business process management software system, decision and agility rules from SPSS software itself, and override rules from the mainframe.

The solution consists of a few different components. First, Santam and SPSS South Africa established a set of business rules in IBM SPSS Risk Control Builder 5 (RCB 5) software, so they could segment data early in the claims process according to parameters such as claim type, amount, and any risk factors that would be red flags for fraud. For example, statistically most car accidents occur at around 10am. But most fraudulent "accidents" are reported between 10pm and 5am. So, the time that the accident is reported is a key factor in determining claim risk. When a claim is submitted, the predictive analytics solution captures data based on these rules. The RCB 5 software scores each claim in real time according to risk factor and IBM SPSS Collaboration and Deployment Services software then segments those claims into established risk categories:

- **Immediate:** Low risk. These claims are settled as quickly as possible, without further assessment.

- **Digital:** Claims that require photographic evidence, which is analyzed and verified by a team at Santam's head office.

- **Mobile:** Claims that are assessed by a Santam operative who visits the site of the claim.

- **Complex:** Claims that require a more sophisticated assessment by an expert team, usually due to legal issues.

- **Merit:** High-risk claims that are considered most likely to be fraudulent. These claims go directly to a special unit for thorough investigation.

By assigning the right team, and thus level of scrutiny, to each claim, Santam was able to find business and customer benefits. One benefit was reducing the amount of time to process low risk claims by nearly 90 percent. Which, while impressive, pales in comparison to saving 2.5 million dollars in payouts to fraudulent customers.

Clearview AI has been effectively using facial recognition technology for law enforcement purposes, providing a revolutionary intelligence platform powered by a facial network of over 40 billion facial images. The company's mission is to help reduce crime, fraud, and risk, create safer communities, ensure secure commerce, and enhance national security and military defense. Clearview AI's facial recognition technology is claimed to be more than 99 percent accurate across all demographics and leverages a database of over 40 billion publicly available images, including those from news media, mugshot websites, public social media, and other open sources.

This extensive database and high accuracy have enabled law enforcement agencies to accelerate investigations, identify individuals, and generate leads that aid in solving both simple and complex crimes, ultimately enhancing public safety and providing justice to victims.

Despite its effectiveness in aiding law enforcement, Clearview AI has faced criticism and legal challenges. The company's database has been declared illegal in several countries, and it has been the target of multiple lawsuits and fines. Critics have raised concerns about the potential privacy implications and the lack of regulations surrounding the use of facial recognition technology by law enforcement. While the technology has been used by police for various crime-solving purposes, including solving murders, the controversy surrounding its use has led to calls for increased accountability and regulation to address potential privacy and ethical issues.

A fascinating book, *Your Face Belongs to Us,* by Kashmir Hill provides detailed background into facial recognition and how Clearview AI and others are using such software to advance fraud detection and other similar applications.

Here are four more examples of how companies have enhanced trust and safety for their customers through innovative practices:

1. **Apple's App Transparency Tracking:** With the release of iOS 14.5, Apple introduced App Tracking Transparency, requiring apps to request permission from users before tracking their activity across other apps and websites. This feature enhances user trust and safety by giving them control over their personal data and how it's shared.
2. **Google's Safe Browsing:** Google offers Safe Browsing technology, which warns users when they attempt to navigate to dangerous sites or download dangerous files. This feature protects users from phishing attempts, malware, and other online threats.

3. **Microsoft's Windows Hello:** Microsoft introduced Windows Hello, a biometric-based technology that allows users to authenticate access to their Windows devices and applications through facial recognition and fingerprint scanning. This feature enhances security and user convenience, fostering trust by ensuring that devices are accessible only by authorized users.
4. **American Express' Use of Big Data for Personalized Service:** American Express analyzes transaction data to provide personalized service recommendations and alerts. This proactive approach to customer service helps prevent fraud and enhances customer experience by using data to anticipate and address customer needs.

These examples showcase how various industries have implemented measures to enhance the trust and safety of their services and products for customers.

We could cover countless other companies who have seen big results from AI. The impact to keeping customers safe and building trust is significant, whether it's a simple action like placing the photo of a technician in an email or running a complex process to predict crime or fraud. The core lesson is one of possibilities. Regardless of your industry, some of these approaches will certainly help you rethink how you approach trust and safety for your customers.

PART V:

THE I.D.E.A. FRAMEWORK

CHAPTER 15

INQUIRE

"A subtle thought that is in error may yet give rise to fruitful inquiry that can establish truths of great value."

- Isaac Asimov

Since the release of *Blue Goldfish* in 2016, we've trained hundreds of companies and organizations. As you might imagine, it's not enough to help an organization understand what a Blue Goldfish is and why they need them, organizations also need help creating their own Blue Goldfish, so we developed the I.D.E.A. Process.

Here's the overview of the I.D.E.A. Process:

Inquire – Understand what is important to your customers and what gaps and opportunities exist.

Design – Generate ideas for Blue Goldfish to address the gaps and opportunities.

Evaluate – Complete an evaluation process to determine which Blue Goldfish ideas should be rolled out across the organization.

Advance – Bring your best ideas out of the pilot phase and into widespread rollout, measuring the results and creating a feedback loop back to the Inquire phase.

In these next four chapters we'll walk you through the steps of the I.D.E.A. Process. Let's get started with Inquire.

Inquire is to ask questions to find direction. During the inquire phase you're going to ask lots of questions.

Before you undertake the creation of a Blue Goldfish, you must first understand the big picture, something customer experience (CX) professionals call the customer journey. By understanding your customer's journeys, and that no two are alike, you can discover the gaps and opportunities to create Blue Goldfish. Let's start by learning more about customer journeys, gaps, and opportunities.

CUSTOMER JOURNEYS

As the name implies a customer journey is everything your customers experience as they interact with your brand before, during, and after purchase. Think for a moment about the last time you used a hotel. Could you list every interaction you had with the hotel before, during, and after your purchase? If you could, your list might look something like this:

- Decide to take a trip
- Research places to stay
- Select an area of town
- Look for hotel options
- Compare hotels
- Select a hotel
- Compare hotel rooms
- Book a hotel room
- Travel to your destination
- Locate your hotel
- Enter the lobby
- Wait for a front desk clerk
- Check in
- Find your room
- Open the door
- Walk inside
- Stay one or more nights
- Check out of your room

- And then travel back home
- Tell others about your experience

This might seem like a long list, but it only scratches the surface of what steps a customer might take in their journey with your company. CX professionals would call each of these steps a touchpoint.

AI has the power to revolutionize the customer experience employing advanced data analytics, predictive analytics, and personalization techniques to identify and enhance key customer touchpoints. By processing vast amounts of data to uncover behavioral patterns and sentiment, AI is uniquely adept at predicting future customer needs, and potential personalized interactions for a tailored experience. AI-driven automation streamlines feedback collection and analysis, ensuring real-time insights across all touchpoints. By detecting friction points, such as navigation issues or checkout barriers, AI facilitates proactive improvements in the customer journey. AI has the potential to see touchpoint issues that may not be obvious to the observer.

CUSTOMER INSIGHTS

There isn't a right way to gather customer insights, but there are plenty of wrong ways. Let's start with the most common one. It sounds something like this: "We've been in business for thirty years. We know exactly what our customers want." Cue the dramatic music. The biggest trap you can fall into is assuming you know exactly what your customers want. Sure, you likely have a good sense of their needs, but there's always so much more you can learn.

There are entire books on gathering customer insights, so we won't try to recreate that here, but what we will offer are some thoughts on the best sources of customer insight. As you prepare to map your customer's journey, you want to answer the following questions:

- What do our customers care about? Which one do they care about most?
- What do customers expect from similar organizations?
- How well do we meet their expectations today?
- What are common points of failure for us?

As you seek to answer these questions, there are numerous sources of input that may be helpful. Here are a few that we suggest our clients turn to:

- Customer data – What data does your organization gather that might provide insights into your customers?
- Surveys – Do you conduct any customer or employee surveys, including Net Promoter or Customer Effort scores?
- External data – Are external data or reports available about your customer segment?
- Interviews – Can you sit down and interview a sample of your customers?
- Win-Loss Analysis – Have you conducted an analysis of customers you didn't win or customers you lost?
- Proxies – If you don't have direct access to customers, who interacts with them, and can you interview them?
- Observations – Can you observe customer interactions either in-person or via recordings?

Often companies will document insights about customers in a document called a persona. Customer personas are profiles of prototypical customers, based on data from actual customers. Often these personas will be organized by the type of customer. For instance, a hotel might have different personas for business travelers, leisure travelers, and event planners, to name a few of their key audiences.

In our workshops, we often create proto-personas. These are persona-like documents that allow our participants to get the feel of creating personas but are often based on their personal experiences and not on customer data. Proto-personas can be helpful for training or for creating a first draft of personas but must be validated by data to be useful.

Personas are great at documenting what your customers want, but often do not include how well you perform against their desires. For that exercise, we often create an Attribute Map. An Attribute Map is a graph where you list your customer's needs on one side in rank order and then rate your relative performance on a scale of 1-5.

An Attribute Map clearly shows what your customers need and where you fall short. This high-level exercise can help you decide where to focus your efforts or can inform the journey mapping process in the next step.

Armed with customer insight, either in its raw form or organized into personas, you now know what your customers want and expect. The Attribute Map and other data about your current performance give you the raw insights that will help you document your customer journey in step two and then identify gaps and opportunities in step three.

INQUIRE PHASE PROGRESS

✓	**Step One:** Gather Customer Insight
	Step Two: Create Journey Maps
	Step Three: Identify Gaps and Opportunities

JOURNEY MAPS

Journey maps are designed to create a sense of empathy. By looking at all your customer touchpoints, good, bad, and ugly, all on one document, you can step into the customer's shoes and feel their pain or delight. Of course, the latter is what we're going for.

In our workshops, we encourage participants to create both current state and future state journey maps. The current state documents what is happening today, while the future state is the journey you might aspire to create.

You can choose to create a high-level journey map that covers all steps of the customer's journey but with less detail or you can choose to create a detailed map for a specific segment of the customer journey to help you perfect a critical part of the journey.

Journey maps do not have to be exhaustive, in fact it's easy to get caught up on creating the most exhaustive journey map possible. We believe this takes your focus away from solving your customers' challenges or worse causes you to miss the proverbial forest for the trees.

Often in our workshops, we ask participants to organize their journey maps in rows by the touchpoint's medium, as different touchpoints happen via social media, via the Web, in-person or, in some cases, through a reseller or agent.

There are several great resources out there on journey mapping, so we won't seek to reinvent that wheel here, but getting back to our hotel example, here's a list of journey maps you might choose to create and how they could be useful.

- **The booking experience** – a detailed journey map including all the steps needed to book a room, divided by channel, which includes social media, the hotel website, the call center, travel agents, and travel booking websites. You would use this map to look for gaps and opportunities to streamline and simplify the booking process.

- **The in-room experience** – a detailed map of the things a customer might choose to do from their guestroom and the related touchpoints. This map could include everything from making coffee, using the iron, or ordering room service. You would use this map to make improvements to your facilities, amenities, or customer service.
- **The high-level experience** – a not-so-detailed map of every phase of the customer's journey with your hotel, with external factors included. This might resemble the list earlier in this chapter. Some external factors might include a flight delay or other travel irritation. You would use this map to help team members have greater empathy for the customer and look for opportunities to improve a customer's trip, even if you didn't cause the underlying issue.

There are all sorts of ways to approach journey mapping and there's no right answer, except that you must consider two factors:

1. Does the journey map give you enough detail to understand how you might improve the customer's experience?
2. Is the journey map based on actual insights from your customers, data about your customers, or insights from your front-line employees?

If the answer to those questions is anything other than a confident yes, you likely need to adjust your approach.

INQUIRE PHASE PROGRESS

✓	**Step One:** Gather Customer Insight
✓	**Step Two:** Create Journey Maps
	Step Three: Identify Gaps and Opportunities

GAPS AND OPPORTUNITIES

A few times during this chapter we've mentioned gaps and opportunities. Now it's time for us to talk more about those.

A gap in your customer journey is a failure or a breakdown. These are points along the journey where your business processes get in the way of the customer's goal or do not serve their needs. The result is often customer frustration.

Unexpected cases of customer frustration can occur as well. Journey mapping helps you minimize these cases, but they always will exist, so create an "issue resolution" journey map to plan how you might handle the occasional misstep or mistake.

But those one-off mistakes aren't what we mean by gaps. Gaps are systemic breakdowns in your customer experience. For instance, at a SaaS company, a journey mapping process revealed a common point of customer frustration. As soon as a contract was signed for a new software subscription, the customer's subscription was activated, and they were billed for setup fees and their first month's fee. Sounds reasonable, yes? But what the journey map revealed was a delay—often by as much as several weeks—before the customer was able to begin using the software. The systemic gap was a customer paying for something they were not yet receiving value from. First impressions matter and this company was not making a good one.

To be clear, we're not suggesting you take the same approach in your organization as your circumstances may be different. What we do hope is that you now understand the difference between a gap, which is a systemic issue, and a one-off issue. It is worth noting, however, that if the same one-off issue keeps coming up, you might want to investigate the root cause to see if it's a systemic issue.

As you review your journey map, think of gaps as valleys or potholes. These are the parts of the customer experience you want to address. Speaking of potholes, let's use a roadway analogy for a moment.

Let's say that your customer journey resembles a road that is flat and straight, and you've pointed out several potholes. Suppose you use all your budget to fill every pothole, but you still have a road that's flat and straight. While there are no issues riding down your road, there's also not much excitement.

Now envision a curvy mountain road—the kind of road where motorists will travel from afar to experience. There are twists, turns, views, and any number of hazards that create a thrill. Let's say that there are also a fair number of potholes.

Which road would you rather experience? The perfectly maintained but boring pavement or the exciting but imperfect one. We suspect it's the latter.

In the *Power of Moments*, Chip and Dan Heath caution against using all your resources to address gaps. They advocate for the creation of peak moments. That is investing in experiences that are memorable and stand out.

Gaps are easy to find and easy to spend all your time chasing. If you do that, however, you'll engage in a never-ending game of Whack-A-Mole and never create an experience that is remarkable. We encourage you to address your largest gaps, but also spend time identifying the opportunities to create a peak moment. Some of the best Blue Goldfish create peak moments for customers.

Identifying opportunities is much more difficult than looking for gaps. Opportunities can exist in a gap. For instance, you could choose to double down on a common gap in your industry not only to solve it but to turn it into a peak experience.

This reminds us of Disney's Be Our Guest restaurant. A common gap in restaurants is wait time. Disney addressed this by creating an order-ahead system so that you spend minimal time waiting for your food at your table. Using their MagicBand system, they've also sought to make this a peak experience where they can deliver your food almost as soon as you are seated (because the RFID in the MagicBand tells them when you arrive and to begin preparing your order). They take the experience to the next level by having servers address you by name without asking your name first. The MagicBand system tells them who's who. And it's especially exciting for kids to experience the "magic" of everyone knowing their name.

For the most part, however, your opportunities won't come from gaps. They will come from ordinary experiences. We ask our workshop participants to review their journey maps for ordinary moments that might resonate with their customers.

For instance, one of the new Blue Goldfish in this 2.0 edition comes from The Durham Hotel in Durham, NC. The retro-chic boutique hotel is a haven for today's modern consumer. The property features a gourmet coffee bar and a lavish rooftop patio, not to mention well-designed, modern guestrooms. Knowing that their hotel attracts a foodie audience, they decided to transform the in-room coffee experience. A coffee pot is a staple of most U.S. hotel rooms. Most hotels purchase filter packs of Starbucks or another recognized coffee brand and ensure rooms are well-stocked. Some hotels take it to the next level with an in-room Keurig and a selection of K-cups. The Durham, however, takes a different approach. Each guestroom has a door hanger that resembles a breakfast room service menu, except that at The Durham the card lists a choice of beans from Durham-based Counter Culture, one of the top five craft roasters in the country. The card allows you to select your beans and the time of day when you would like for your freshly ground coffee to be delivered. Oh, and did we mention that the coffee service is complimentary? In the morning, by your chosen time, you'll have a sealed packet of your freshly-ground coffee choice waiting for you outside your door. Talk about a "perk."

Before creating this experience, we're certain that The Durham didn't see coffee as a gap. Customers are generally not expecting a gourmet, in-room coffee experience, but it was an opportunity to create a peak moment.

INQUIRE PHASE PROGRESS

✓	**Step One:** Gather Customer Insight
✓	**Step Two:** Create Journey Maps
✓	**Step Three:** Identify Gaps and Opportunities

Now that we've completed the Inquiry stage, it's time to start creating Blue Goldfish. Let's move into the Design phase of the I.D.E.A. Process.

CHAPTER 16

DESIGN

"Design is not just what it looks like and feels like. Design is how it works."

- Steve Jobs

Design is the most exciting part of the I.D.E.A. Process. It's where you generate ideas, some of which may very well be innovative, to better serve your customers. Ideation is rarely a linear process. Some ideas come from an epiphany and others build on each other.

While ideas come to us when we least expect it, there are ways to organize the ideation process and that's exactly what we recommend during the Design phase. This phase calls for you to brainstorm your Blue Goldfish around the gaps and opportunities you identified during the Inquire phase.

The goal during the design phase is to generate as many ideas as possible, whether or not you believe they will be immediately viable. Often the best ideas come from creating a more reasonable or measured version of an idea that's too big, complicated, or expensive to implement.

SET YOUR FOCUS

As we discussed in the previous section, your goal should not be to fill every gap you identified. Instead, your goal is to address the gaps that are most damaging to your customer experience. In an ideal world, this will leave budget, time, and other resources to create peak moments for the most promising opportunities.

To this end, your first step in the Design phase is to choose which gaps and opportunities you would like to address. Starting from the list you created, first identify which gaps are the most damaging. Here are some questions you can ask to help you prioritize the list:

After which gaps do, we tend to ...

- lose the most customers?
- see NPS or other measures drop?
- receive the most complaints?

As you ask these questions and rank the gaps accordingly, chances are good that a few will bubble to the top. Those 3-5 top gaps are the ones you'll want to design for.

We also encourage you to repeat this ranking process for opportunities but using a different set of questions to determine the most promising ones. Here are questions to get started:

Which opportunities...

- are directly related to your core product or service?
- are most important to your target customers?
- will serve most of your customers or prospects?
- will create competitive separation in the marketplace?
- are unlike anything we've done before?

As you can imagine, ranking opportunities is a bit harder. The process requires intuition and will feel more squishy than ranking gaps. This is exactly why so many companies fall into the trap of only addressing gaps. Our rational brains far prefer the certainty of addressing known gaps to the uncertainty of creating something brand new. Embrace the uncertainty.

At this point add your 3-5 opportunities to your list of 3-5 gaps. These are your areas of focus for the Design process. You might have completed this ranking alone or with a group. If you worked alone, now is the time to get others involved.

DESIGN PHASE PROGRESS

✓	**Step One:** Set Your Focus
	Step Two: Ask Big Questions
	Step Three: Organize Your Ideas

ASK BIG QUESTIONS

Just before the brainstorming portion of our workshops we ask participants to complete an exercise. We give each person a blank sheet of paper and ask them to draw a picture of a vase with flowers. Within about 30 seconds, most everyone, regardless of artistic talent, sketches out a container with one or more flowers. We then ask them to draw a second picture. This time, however, we ask them to draw a better way to enjoy flowers in their home. As you might imagine, we receive many different drawings. Often, we go around the room and ask each participant to explain their idea.

We call this a design thinking exercise. Psychologically speaking, it's an example of divergent and convergent thinking. Both terms were coined by psychologist Joy Paul Guilford.

Divergent thinking is a thought process or method used to generate creative ideas by exploring many possible solutions.[82]

Convergent thinking is the opposite of divergent thinking. It generally means the ability to give the "correct" answer to standard questions that do not require significant creativity.[83]

During the exercise, the first prompt—drawing a vase—is a convergent thinking exercise. While some creativity is involved, there is generally one correct answer. The second prompt requires divergent thinking as we're asking participants to come up with a possible solution to the problem and not one correct answer.

You might be wondering what this has to do with designing your Blue Goldfish. Everything, in fact. This exercise exists to awaken the parts of our brains that handle divergent thinking. As we start the brainstorming process, we're not looking for the most correct

82. https://en.wikipedia.org/wiki/Divergent_thinking
83. https://en.wikipedia.org/wiki/Convergent_thinking

or logical ideas. Rather we are seeking many possible solutions to address the given gap or opportunity.

Harvard Instructor Anne Manning notes that divergent thinking is useful for coming up with ideas and convergent thinking is good for making decisions about those ideas, but it's difficult to do both at the same time.[84] During the Design phase, you and your team should focus on divergent thinking and avoid the temptation to question or evaluate ideas.

We recommend a group brainstorm for generating ideas. There are whole books on the art and science of brainstorming, so we'll just cover our tips for the process.

1. **Environment**. Create an environment for generating new ideas. You might want to consider hosting the brainstorm at an off-site location or perhaps immediately after touring one of your retail locations.
2. **Supplies**. Make sure everyone has what they need to be productive—coffee, water, snacks, office supplies.
3. **People**. Get the right number of people in the room. Brainstorming works best with 5-8 people. Beyond that number, some will mentally check out of the process.
4. **Cross-pollination**. In our workshops, participants are often grouped with others from their teams in the company. This can work well as can intentionally mixing departments to generate different perspectives. We like and use both methods and encourage you to experiment with what works best for you.
5. **Focus and timing**. We recommend focusing on one gap or opportunity at a time. Perhaps even setting a timer or a goal

84. https://www.youtube.com/watch?v=xjE2RV6IQzo

for total ideas and once completed, moving on to the next gap or opportunity.

6. **Format**. In our workshops, we often ask for individuals to first come up with ideas and then share them with their brainstorming group. Sometimes we use sticky notes to collect all of the individual ideas.
7. **Yes, and**. Remember, the purpose of the exercise is to generate as many ideas as possible, not to evaluate them. You'll want to set strict rules about not dismissing or contradicting any ideas. A helpful way to do that is to start ideas with "yes, and" while avoiding words like "but" and "however" (and other creative phrases to contradict without using these words.)

We called this section "Ask Big Questions" and now that we're set up to brainstorm, we finally get to ask those big questions. We find that it's helpful to ask thought-provoking questions around each gap or opportunity to elicit different ideas to address them. During the brainstorm, your facilitator may choose to ask some of these questions to help elicit ideas.

Here's our go-to list of questions:

- If it were magic, how would it happen?
- If we eliminated x altogether, what could we replace it with?
- What is the confusing part about x?
- How have others solved x?
- If you had a budget of one million dollars, how would you solve this problem? What about a budget of ten dollars?
- If you had a year to solve this problem, how would you do it? What if you had ten minutes?

You will likely develop some of your own questions as well. We like questions that go to the extremes, as you can see with the last two questions focused on budget and timeline. Often the best ideas come from the pared back version of a much larger, but infeasible idea. Those questions help develop those big ideas.

As you finish the brainstorming process, make sure you've captured all your ideas. Taking photos of whiteboards, sticky notes, and flip-charts can be helpful. These will be important reference materials as you move into the next step.

DESIGN PHASE PROGRESS

✓	**Step One:** Set Your Focus
✓	**Step Two:** Ask Big Questions
	Step Three: Organize Your Ideas

ORGANIZE YOUR IDEAS

Now that you've generated many ideas, some of which are likely extraordinary, you'll want to organize them in a uniform way for internal consumption and evaluation. There's no right way to undertake this process, but we do have a few recommendations.

Start by stating the gap or opportunity you're seeking to address. Then, include each idea. For the idea, you'll likely need a short title and a longer description to explain it. For instance, you might title an idea "Billing Accuracy" and describe it as "offer a guarantee to customers that our bills will be accurate, or they will receive a refund plus 50 percent of the error." This will allow you to talk about the idea before building it out.

Broadly, your ideas will fall into the categories of value and maintenance. Generally, but not always, ideas to fill gaps will seek to reduce maintenance and ideas to address opportunities will seek to increase value.

You may find it helpful to organize your ideas into the Blue Goldfish framework at this point. It's certainly not required but it can help you to compare your ideas to those Blue Goldfish that already exist in the wild.

At this stage, you now have an organized list of gaps and opportunities, along with a list of ideas—the Blue Goldfish—that you can now evaluate. Of course, generating ideas is the easy part. Implementation is much harder.

DESIGN PHASE PROGRESS

✓	**Step One:** Set Your Focus
✓	**Step Two:** Ask Big Questions
✓	**Step Three:** Organize Your Ideas

WHERE AI AND CONTROVERSY COLLIDE

The human element in our I.D.E.A. process is critical. Yet, artificial intelligence has the potential to significantly impact the process of problem-solving, design, evaluation, and advancement. The integration of AI in these processes can lead to rapid advancements and enhanced outcomes. AI can support divergent thinking, aid in the generation of novel ideas, and help overcome challenges in evaluating and synthesizing different ideas. Additionally, AI can provide adaptivity at the step-by-step level and at scale, making it a valuable tool in the design and problem-solving process. However, human intervention remains crucial in various aspects of the process, and the unique capabilities of human creativity and critical thinking are irreplaceable.

AI's Role in Problem-Solving and Design

AI can be leveraged to supplement and support the concept of creating insight into a problem, designing solutions, evaluating outcomes, and advancing the process. It can assist in promoting divergent thinking by making associations among remote concepts and producing novel ideas. Furthermore, AI-enabled search platforms can help designers discover inspirational stimuli to aid in the solution-finding process. The integration of AI in problem-solving and design processes can lead to more user-centered and continuously updated solutions, overcoming past limitations in scale, scope, and learning.

Human Intervention and AI's Advantages

While AI can expedite the problem-solving and design processes, human intervention remains essential. Human creativity, critical thinking, and the ability to understand complex contextual nuances are irreplaceable. AI can excel in tasks that involve large-scale data processing, pattern recognition, and providing adaptivity at the step-by-step level. It can also help overcome challenges in evaluating and synthesizing different ideas, making it a valuable tool in the creative process.

The Synergy of Human and AI Capabilities

The synergy of human and AI capabilities holds the potential for groundbreaking advancements in problem-solving and design. By leveraging AI to handle repetitive and data-intensive tasks, humans can focus on higher-level ideation, creativity, and critical decision-making. This collaborative approach can lead to more efficient and innovative solutions, ultimately driving progress across various domains. In conclusion, the integration of AI in problem-solving and design processes has the potential to drive rapid advancements and enhance outcomes. While AI can support divergent thinking and provide adaptivity at scale, human intervention remains crucial for complex contextual understanding, creativity, and critical decision-making. The synergy of human and AI capabilities holds the key to unlocking new frontiers in innovation and problem-solving.

As we move forward let's embrace the possibility that AI, especially the large language models (LLMs) will be beneficial to making the Blue Goldfish I.D.E.A. process easier, more effective, and allow for human creativity when finding solutions.

CHAPTER 17

EVALUATE

"To prepare for the future, we must be willing to test new concepts.
This means we must acquire enough information to evaluate these concepts and not be like travelers in a foreign land who compare everything with their hometown."

- Jacque Franc

Evaluate is the newest part of the I.D.E.A. process. The previous model—the 3Ds of Discover, Design, and Deploy—didn't have enough emphasis on testing ideas before rolling them out. It's not that we didn't advocate for testing before, rather it's that we believe it is now more important than ever.

During the evaluation process, we recommend both internal and external validation. Of course, most of the Blue Goldfish Fry—the ideas—came from people inside your company. Along with the valuable insight each employee brought to the idea came the not-so-valuable biases they inherently have.

In the previous chapter we discussed the concepts of divergent and convergent thinking. At this point in the process, we want to re-awaken your convergent thinking ability as our objective is no longer to generate ideas, but to determine analytically the viability of each.

With the list of Blue Goldfish Fry (BGF) you generated in the Design phase, you'll now dive into Internal Evaluation.

INTERNAL EVALUATION

Your internal evaluation phase starts with the list of BGF you're considering. You'll likely have somewhere between six and ten ideas. You'll want to determine if each of them is likely to work. While we're not yet involving customers, which we'll get to in the next two sections, you'll use this time to evaluate each BGF viability from the company perspective.

Is the idea feasible? Like many of the concepts we discuss, feasibility studies can be quite complex. We recommend that you use your own judgment as to how thorough you would like for your feasibility study to be. If you're a small business, we recognize that this process might be simple. For a large enterprise, we realize this stage is often where good ideas get stuck. Here are the questions you'll need to answer:

- What resources will be needed to implement this idea? What are the costs associated with those resources?
- Do the costs outweigh the benefits?
- If you're successful, how many of your current customers are likely to take advantage of this Blue Goldfish? Do you expect increased loyalty from these customers?
- If you're successful, how many new customers do you expect this Blue Goldfish may attract?
- Can you expect benefits beyond additional sales and loyalty? Are there any intangible benefits?

It's tempting to think of internal evaluation as an elimination round, but we like to think of it as a checkpoint instead. Instead of looking for reasons to eliminate a Blue Goldfish from consideration, you're looking for reasons to advance it forward. On the other hand, if you find information that leads you to believe the idea is not viable, you have three options. You can either eliminate the Blue Goldfish, adjust it to be viable, or wait until after external validation to make a final decision.

In either case, you have now evaluated the feasibility of your Blue Goldfish. We hope your excitement is growing, because the next step is to start getting input from the most important evaluators—your customers.

EVALUATE PHASE PROGRESS

✓	**Step One:** Internal Evaluation
	Step Two: External Validation
	Step Three: Pilot

EXTERNAL VALIDATION

At this stage, we advocate for two types of input from your customers. The first we call validation. The idea behind validation is use an inexpensive approach for feasibility and avoid the cost of a pilot in the event the idea does not resonate with your customers.

External validation resembles traditional market research. You'll share your ideas with your target audience and receive their feedback. There's no one correct way to accomplish this, but there are plenty of incorrect ways. Here are some of the ways we recommend seeking input from your customer.

- Focus groups
- Customer advisory board meetings
- Surveys
- One-on-one customer interviews

No matter which research method you choose, you will want to make sure you do a few things. First you will want to make sure you describe your idea fully to the customer. The title and short description you developed in the Design phase will be helpful, but you may want to include more detail at this stage.

Second you should get the customers' initial reaction to the idea and then ask if they have any questions about it. First impressions only happen once, so you'll want to capture them immediately.

Third, you'll want to engage further with the customers, asking them what they like and dislike about the idea and if implementation of the idea would cause them to do more business with you. Of course, a positive response is not a guarantee as researchers will tell you. Humans are bad at predicting their future behavior.

Like many of our recommendations, whole books (and dissertations, for that matter) have been written on market research. You may want to pursue DIY research, or you may want to engage a professional researcher or firm. Engaging a researcher or firm will be more expensive, but it may make sense if you have the budget, and the stakes are high. If you choose the DIY route, we encourage you to seek out additional resources on market research to guide you.

Once your research is complete, you will need to make decisions about each one of your Blue Goldfish. You have a few options:

1. Move the Blue Goldfish into a pilot study
2. Revise the Blue Goldfish and complete more research
3. Hold the Blue Goldfish for a future pilot study
4. Eliminate the Blue Goldfish

You probably will have a mix of all the above for your project and that's perfectly okay! It's why we do research before rolling out a pilot project. This research most likely generated some additional insights about your customers that you will want to consider during your pilot, or it may cause you to revisit your journey maps with fresh data.

Jack Welch famously said, "An organization's ability to learn, and translate that learning into action rapidly, is the ultimate competitive advantage." Now that you're armed with customer input, you're ready to start putting your ideas into action.

EVALUATE PHASE PROGRESS

✓	**Step One:** Internal Evaluation
✓	**Step Two:** External Validation
	Step Three: Pilot

PILOT

We're excited you've made it to the customer-facing pilot! There's only so much you can learn from the process we've outlined here. Now you get the privilege of learning directly from your customers in a real-world scenario. We are legitimately excited for you!

Pilot programs differ depending upon the size of the organization and customer base. Your pilot may be as simple as trying out one Blue Goldfish for a day or as complex as rolling out the Blue Goldfish in a limited number of your locations for a longer period. You're the best judge of what will work most effectively for your customers and organization.

Here are some tips we recommend you keep in mind as you put together your pilot:

- Make sure your pilot is sufficiently sized, by reach or timeframe, to generate a representative sample of your target customers. Keep in mind that your Blue Goldfish may be targeting a subset of your customers, so you'll want to consider the locations and or timeframes when those customers are most likely to engage.
- Build measurement into the process. As you are planning the pilot, you will want to determine how you'll measure the effects of the Blue Goldfish. An uptick in sales would be marvelous, but sales may be a lagging indicator. Consider other ways to measure those customers or locations that are part of the pilot relative to the rest of your customers. Ideally you would be able to segment the pilot customers from other customers and compare multiple metrics like survey results, sales, order value/volume or NPS.
- Make the experience as real as possible. We suggest not telling the customer you are *trying out* something new. Simply say it's something new. In physics, there's a

concept called Observer Effect. That is the theory that the mere observation of a phenomenon inevitably changes that phenomenon. We're not physicists, but we do suspect that the knowledge that something is a pilot will change a customer's perception of it.

- Engage in proper training for your team. Remember when outward facing, it's not a pilot, it's the new thing. You'll want your team to treat the Blue Goldfish pilot as normally as possible to avoid skewed results or apathy toward the project.
- Listen to your team. To achieve buy-in and an effective launch, you'll also want to seek the feedback of your team members tasked with delivering your new Blue Goldfish experience. This input will allow you to head off any opposition or issues that might interrupt an effective roll-out.

Once you deploy the Blue Goldfish pilot and measure its results, you arrive at your final point of analysis—the go or no-go decision. Next, you'll prepare to roll out any Blue Goldfish that produces desirable results more widely in your organization. It's Blue Goldfish graduation day as they're now qualified for deployment.

Let's turn our attention now to helping you roll them out successfully!

EVALUATE PHASE PROGRESS

✓	**Step One:** Internal Evaluation
✓	**Step Two:** External Validation
✓	**Step Three:** Pilot

CHAPTER 18

ADVANCE

"The advance of technology is based on making it fit in so that you don't really even notice it, so it's part of everyday life."

- Bill Gates

Welcome to the last step of the I.D.E.A. process—the **Advance** stage. Advancing your newly commissioned Blue Goldfish isn't a one-time event. It's a continuous process—one where you seek to advance the Blue Goldfish and the overall experience you create for your customers to a new level.

During the advance stage, you have three primary responsibilities:

1. Achieve buy-in internally
2. Roll out the experience to customers
3. Set up a feedback loop with continuous measurement

BUY-IN

Every organization operates differently, and your buy-in process will differ accordingly. As a small company, you may only need to achieve buy-in from a small group. For larger companies, you may have a formal review process and a need to achieve approval from several stakeholders. As with considering or implementing any of our advice, you know your organization best, so we're not able to be prescriptive about what buy-in you will need to achieve.

We do believe, however, that there are common components of achieving buy-in, so we recommend that you take all of them into account:

- Communicate the plan to the right stakeholders. You'll want to make sure that everyone who has influence over the project is on-board. Don't forget the influence that teams like operations and finance can weld—involve them as early as possible and ensure you address their concerns.
- Don't forget your front-line employees. These are the people tasked with implementing the Blue Goldfish, so their understanding of both why and how to deploy the Blue Goldfish is essential.

- Present the story of the idea. It might be tempting to say, "We decided to do this, and these are the results." While that is true, a significant amount or work went into formulating the Blue Goldfish throughout the Inquire, Design, and Evaluate phases. Make sure everyone knows how the Blue Goldfish came to be.
- Use anecdotes to win hearts. During the pilot, you no doubt heard anecdotal feedback of how the Blue Goldfish helped a customer. By sharing some of that feedback, you will connect with and engender empathy (and usually support) from your stakeholders.
- Use data to win minds. Thinking back to the discussion about convergent thinking, you know you'll need numbers to prove that the Blue Goldfish was successful. You gathered significant data from the pilot, so use it to share (and prove) what you learned.
- Present adjustments based on pilot feedback. A story where everything is perfect not only isn't believable, but it's also boring. You'll want to present some of the learnings from the pilot as well as the adjustments they necessitated to solidify buy-in.

We can't guarantee an easy buy-in process if you follow these tips, but we can guarantee that you'll set yourself up for success. Once you achieve buy-in, the most fun part comes next—rolling your Blue Goldfish out to customers!

ADVANCE PHASE PROGRESS

✓	Step One: Buy-In
	Step Two: Rollout
	Step Three: Measurement

ROLLOUT

Let's do this! Rollout is one of the most exciting and critical parts of the journey. You'll soon see the fruits of your labor, but to do so you'll need a successful roll out. Just as in each of the previous phases, how you roll out your Blue Goldfish will be based on your organization, the resources involved, and any number of other factors. Much as before, this leaves us to offer our generalized guidance. We encourage you to adapt as necessary to suit your needs.

Here are some of the factors you should consider during roll out.

- Communication – How will you communicate the Blue Goldfish offering to both front-line employees and customers?
- Policies – How clear are your policies for what's allowed and disallowed? Have you considered edge cases or exceptions?
- Logistics – What new materials need to be acquired and distributed to the front-line?
- Metrics – How will you collect relevant metrics? You measured during the pilot, so you already have a start. How will standard measurement differ from the pilot?
- Promotion – As you know from reading this book, we don't advocate for Blue Goldfish as a PR strategy, but you may need to promote the offering to gain initial usage from customers.

Rollout may seem simple; however, it is anything but. How well you roll out your Blue Goldfish may determine its fate. This time might be marked by late nights and early mornings, but all of the work you've done up to this point should provide the necessary motivation to make it happen. We believe in you (and your Blue Goldfish)!

ADVANCE PHASE PROGRESS

✓	Step One: Buy-In
✓	Step Two: Rollout
	Step Three: Measurement

MEASUREMENT

As you know from the pilot, measurement is how you prove your Blue Goldfish is working and delivering the desired results for the business. Measurement is also how you gain valuable insights that will help you return to the Inquire phase and address new (or different) gaps and opportunities. Here are some of our tips for effective measurement, realizing, as you know by now, that every organization is different.

- Be sure to collect performance data. You identified the metrics as a part of the roll out, so now you'll want to collect and analyze the data.
- Use segmentation to understand relative performance. Over time the enthusiastic buy-in you achieved may wane, so you'll want to show how the Blue Goldfish consistently delivers results. Ideally, you will be able to compare key metrics between those customers who experience the Blue Goldfish and those who did not.
- Mind the bottom line. You'll want to connect the performance of the Blue Goldfish to metrics as close to the bottom line as possible. By showing increased loyalty and profitability because of the Blue Goldfish, you're likely to

retain the necessary support to keep the goldfish swimming. (No apologies for the pun.)

- Refine as needed. There's no reason you can't tweak as you go along. Despite the best planning and analysis, your Blue Goldfish may miss the mark slightly. Don't hesitate to adjust as needed but resist the urge to make them too early. There's no right answer here, so follow your gut. It's usually right.
- Build the feedback loop. Continue collecting the necessary data to loop back to the Inquire phase. As you improve, the desire to improve will grow in your organization. Use this desire, your new data, and, hopefully, your new profits to fuel the feedback loop. This will create a flywheel effect that allows your organization to reach new heights on a continuous basis.

Woohoo! You've made it through the I.D.E.A. Process! Our excitement level is high, because we know that with every Blue Goldfish out in the wild, there's a customer somewhere whose experience is just a little better than it was before. It's why we do what we do, and we hope you share the same excitement.

ADVANCE PHASE PROGRESS

✓	**Step One: Buy-In**
✓	**Step Two: Rollout**
✓	**Step Three: Measurement**

WHAT HAPPENS NEXT?

Your goal is to consistently move to the next plateau on each iteration through the I.D.E.A. Process. A Blue Goldfish isn't a campaign—it's a commitment. What was once a differentiator can quickly become an expectation, so consistent improvement is essential in today's hyper-competitive marketplace. New successes come with increased expectations which bring up new issues. It's not just about a completing the I.D.E.A. Process again, it's about consistently serving the customer.

PART VI:

FINAL THOUGHTS

CHAPTER 19

TOP 10 KEY TAKEAWAYS

"Advice is like a tablet of aspirin.
It tends to only work if you take it."

– David Murphy

We've reached the end of *Blue Goldfish 2.0*. Here are 10 key takeaways that sum up the book:

1. Technology has paved the way for every company to know more about their customers and build deeper relationships with them. Especially with the advent of access to AI by substantially all parties, the ability to find effective and creative solutions to problems has been enhanced exponentially. Today's businesses must use data and AI to understand and meet customer needs more effectively.
2. Customer expectations are binary. You either exceed them or you fall short. Similarly, if you're trying to keep up with increasing consumer expectations, you're fighting an uphill battle. Empowered customers are starting to take for granted that a company will know and understand their individual needs. The bar has raised by companies like Amazon, highlighting the need for businesses to exceed expectations by being more responsive, personalized, and anticipatory of customer needs.
3. There are only two sources of competitive advantage: the ability to learn more about our customers faster than the competition (today assisted with the help of AI) and the ability to turn learning into action faster than the competition. Building and maintaining strong relationships with customers through understanding their needs and preferences is fundamental to business success in the digital age.
4. The overarching lesson of data myning for customer experience is empowerment. We believe companies who use personal data to empower their customers have an opportunity to create transformative change for those customers. With empowerment and then success, gratitude and loyalty are often the next set of emotions.

5. To personalize is to design or produce something to meet someone's individual requirements. The key is the individual. Stymied by legacy technology and distributed workforces, personalization can be difficult. Yet the companies who use AI to overtake those barriers tend to see impressive results.
6. Companies need to start delivering a next level of customer service. A level we call 3.0. Data from your own database is gold and the ability to extract vast data related to your customers via AI is not a futuristic concept, but reality.
7. The more you know and the faster you can respond allows you to create differentiation. Sometimes it's surprising and delighting a customer with speed. Other times it's identifying and correcting an issue before a customer even knows about it. And when necessary, it's responding to correct a bad situation before it gets worse.
8. Differentiation is key. Just like goldfish can grow based on their environment and genetic makeup, businesses must differentiate themselves to grow in their market. This differentiation can be achieved through leveraging AI and data to create a unique customer experience.
9. The importance of merging online and offline experiences. Merging online data with offline interactions can create a seamless and highly personalized customer journey, enhancing satisfaction and loyalty.
10. The role of AI in business growth. AI and data analytics are crucial tools for businesses to understand their customers better and act with greater agility, thus enabling personalized experiences at scale and fostering business growth.

These takeaways highlight the book's focus on deepening customer relationships, differentiating from competitors, and adapting to the heightened expectations of today's consumers.

ABOUT THE AUTHORS

STAN PHELPS

Stan Phelps is an author, speaker, and experience architect. He believes that today's organizations must focus on meaningful differentiation to win the hearts of both employees and customers.

He is the Founder of StanPhelpsSpeaks.com, a training company that focuses on customer experience and employee engagement solutions. He believes that *"differentiation via added value"* can be a game changing strategy. For far too long, the overwhelming majority of marketing has fixated on the eyes and ears of the prospect. Not enough has been focused on creating experiences for current customers that drive referrals. Great customer experience is about being so remarkable that people can't help but talk about you. That if you absolutely delight someone – they will not only come back, but they'll bring their friends.

Prior to his writing/speaking, Stan had a 20-year career in marketing that included leadership positions at IMG, adidas, PGA Exhibitions, and Synergy. At Synergy, he worked on award-winning experiential programs for top brands such as KFC, NASCAR, Starbucks, and M&M's.

Stan is the author of 20 books and is a former Forbes contributor. His writing has been syndicated on top sites such as Customer Think and Business2Community. He has spoken at over 750 events in the US, Canada, Mexico, UK, France, Sweden, Honduras,

Holland, Russia, Peru, Israel, Bahrain, Cambodia, The Philippines, Denmark, Germany, Spain, Poland, Japan, Malaysia, Ecuador, Singapore, and Australia.

Stan received a BS in Marketing and Human Resources from Marist College, a JD/MBA from Villanova University, and a certificate for Achieving Breakthrough Service from Harvard Business School. He is a Certified Net Promoter Associate and has taught as an adjunct professor of marketing at NYU and Manhattanville College. He is the Past President of the AMA Triangle, a chapter of the American Marketing Association and NSA Carolinas, a chapter of the National Speakers Association. Stan lives in Cary, North Carolina with his wife Jennifer and two sons Thomas & James.

To learn more about Stan's journey, his perspectives on differentiated experience (DX), or to engage him for speaking or training opportunities, visit stanphelps.com.

CHUCK GALLAGHER

Chuck Gallagher's journey is a testament to the transformative power of ethical decision-making. Starting his career with humble beginnings, Chuck quickly climbed the corporate ladder, only to find himself facing the consequences of unethical choices. This pivotal moment marked a profound ethical awakening, leading Chuck to dedicate his life to educating others about the importance of integrity and ethical decision-making in business.

Emerging as a renowned speaker and author, Chuck Gallagher has become a pivotal figure at the intersection of ethics and artificial intelligence. His insights stem from a deep understanding of the ethical dilemmas posed by advancing technology. Chuck's ability to dissect complex ethical issues and present them in an accessible manner makes him a sought-after consultant for businesses striving to navigate the ethical quandaries of AI integration.

As an author, Chuck's publications resonate with a wide audience, blending personal anecdotes with practical advice. His speaking engagements are not mere lectures; they are interactive experiences, empowering attendees to rethink their approach to ethics in both their professional and personal lives. Chuck's ability to connect with his audience, whether through keynote speeches or intimate workshops, is unparalleled. He is the author of four books as he joins as co-author of this one.

In his consulting work, Chuck doesn't just address the symptoms of ethical challenges; he delves into their root causes, offering bespoke solutions that align with each organization's unique culture and objectives. His approach is not about imposing standards but about fostering an environment where ethical decision-making thrives.

Visit chuckgallagher.com to learn more about Chuck's journey, his perspectives on ethics and AI, or to engage him for speaking or consulting opportunities. Embrace the opportunity to transform your understanding of ethics in the AI-driven world.

ADDITIONAL INSPIRATION AND FURTHER READING

PURPLE GOLDFISH 2.0 – 10 WAYS TO ATTRACT RAVING CUSTOMERS

Purple Goldfish is based on the Purple Goldfish Project, a crowd-sourcing effort that collected more than 1,001 examples of signature-added value. The book draws inspiration from the concept of lagniappe, providing 10 practical strategies for winning the hearts of customers and influencing positive word of mouth.

GREEN GOLDFISH 2.0 – 15 KEYS TO DRIVING EMPLOYEE ENGAGEMENT

Green Goldfish is based on the simple premise that happy engaged employees create happy enthused customers. The book focuses on 15 different ways to drive employee engagement and reinforce a strong corporate culture.

GOLDEN GOLDFISH – THE VITAL FEW

Golden Goldfish examines the importance of your top 20 percent of customers and employees. The book showcases nine ways to drive loyalty and retention with these two critical groups.

RED GOLDFISH - MOTIVATING SALES AND LOYALTY THROUGH SHARED PASSION AND PURPOSE

Purpose is changing the way we work and how customers choose business partners. It is driving loyalty, and it's on its way to becoming the ultimate differentiator in business. *Red Goldfish* shares cutting edge examples and reveals the eight ways businesses can embrace purpose that drives employee engagement, fuels the bottom line, and makes an impact on the lives of those it serves.

PURPLE GOLDFISH SERVICE EDITION - 12 WAYS HOTELS, RESTAURANTS, AND AIRLINES WIN THE RIGHT CUSTOMERS

Purple Goldfish Service Edition is about differentiation via added value and marketing to your existing customers via G.L.U.E. (**g**iving **l**ittle **u**nexpected **e**xtras). Packed with over 100 examples, the book focuses on the 12 ways to do the "little extras" to improve the customer experience for restaurants, hotels, and airlines. The end-result is increased sales, happier customers, and positive word of mouth.

PINK GOLDFISH 2.0 - DEFY NORMAL AND EXPLOIT IMPERFECTION

Companies need to stand out in a crowded marketplace, but true differentiation is increasingly rare. Based on over 350 case studies, *Pink Goldfish 2.0* provides an unconventional eight-part F.L.A.W.S.O.M.E. framework for achieving competitive separation by embracing flaws instead of fixing them.

PURPLE GOLDFISH FRANCHISE EDITION - THE ULTIMATE S.Y.S.T.E.M. FOR FRANCHISORS AND FRANCHISEES

Packed with over 100 best-practice examples, *Purple Goldfish Franchise Edition* focuses on the six keys to creating a successful franchise

S.Y.S.T.E.M. and a dozen ways to create a signature customer experience.

YELLOW GOLDFISH - NINE WAYS TO DRIVE HAPPINESS IN BUSINESS FOR GROWTH, PRODUCTIVITY, AND PROSPERITY

There should only be one success metric in business and that's happiness. A Yellow Goldfish is any time a business does a little extra to contribute to the happiness of its customers, employees, or society. Based on nearly 300 case studies, *Yellow Goldfish* provides a nine-part framework for happiness-driven growth, productivity, and prosperity in business.

GRAY GOLDFISH - NAVIGATING THE GRAY AREAS TO SUCCESSFULLY LEAD EVERY GENERATION

How do you successfully lead the five generations in today's workforce? You need tools to navigate. Filled with over 100 case studies and the Generational Matrix, *Gray Goldfish* provides the definitive map for leaders to follow as they recruit, train, manage, and inspire across the generations.

RED GOLDFISH NONPROFIT EDITION - HOW THE BEST NONPROFITS LEVERAGE THEIR PURPOSE TO INCREASE ENGAGEMENT AND IMPACT

The competition is fierce in the nonprofit world, even when competing in different spaces. This book explores the signature ways nonprofits reinforce their purpose and stand out in a crowded marketplace, whether it is an extra level of recognition for key donors, a special incentive designed to keep their best employees, or something simple like a luncheon to recognize volunteers or highest fundraisers. If you work at a nonprofit, this book will help you deliver "a little extra" to your stakeholders.

DIAMOND GOLDFISH - EXCEL UNDER PRESSURE & THRIVE IN THE GAME OF BUSINESS

Diamond Goldfish uncovers how business is a game. It's a guide for driving sales and deepening client relationships. Based on the Diamond Rule, over 150 case studies, and the science-backed framework of Market Force, the book provides perspective and tools for winning in sales and client management.

RED GOLDFISH PROMO EDITION - HOW PROMOTIONAL PRODUCTS LEVERAGE PURPOSE AND CREATE IMPACT

Red Goldfish Promo Edition is based on a collection of more than 250 examples. The book shares the five ways businesses in the Promotional Marketing industry are tapping into purpose to stand out in business and spread goodness into the world.

SILVER GOLDFISH - LOUD & CLEAR: THE 10 KEYS TO DELIVERING MEMORABLE BUSINESS PRESENTATIONS

How do you avoid giving the typical boring corporate presentation? You need the tools and the approach to delivering memorable presentations. Filled with 64 tips, *Silver Goldfish* provides 10 keys and a six-step approach to coming across Loud & Clear when presenting.

BLACK GOLDFISH - 10 KEYS TO CREATING A DIFFERENTIATED EXPERIENCE

Your brand today is no longer what you tell people it is. It is the differentiated experience (DX) your employees deliver. It is what you stand for and how your customers feel about you. And most importantly, your brand is what your customers and employees tell others about their experience. *Black Goldfish* is an amalgam of the first 10 colors in the *Goldfish Series.*

Made in the USA
Columbia, SC
22 May 2024

35613280R00127